Communicating Science Clearly

This unique self-help guide equips undergraduates, postgraduate students, and early career researchers within the sciences with transferrable communication skills that they can adapt and refer back to as they progress through their careers.

It provides practical guidance on how to best communicate science in a range of different settings. This book facilitates clear and concise communication in both academic scenarios and the workplace. It covers settings ranging from job interviews, through conference presentations, to explaining your research to the general public.

It is illustrated with first-hand experiences, top tips, and "dos and don'ts" to demonstrate best practices. It will also be a valuable guide for established academics who would like a refresher or a guide to new avenues of science communication, such as podcasts.

Key Features:

- Written by an award-winning professional science journalist and broadcaster with 25 years' experience, including writing for national newspapers, devising and presenting programmes for BBC Radio 4, and being interviewed on radio, TV, video, and podcasts
- Covers science communication in a broad range of settings including peer-to-peer, to your manager, at job interviews, and during media appearances
- Includes advice from a range of experts who communicate professionally including a radio producer, a TV presenter, actors and entertainers, and scientists

Communicating
Science Clearly
A Self-Help Guide for
Students and Researchers

Sharon Ann Holgate

CRC Press
Taylor & Francis Group
Boca Raton London New York

CRC Press is an imprint of the
Taylor & Francis Group, an **informa** business

First edition published 2024
by CRC Press
2385 Executive Center Drive, Suite 320, Boca Raton, FL 33431

and by CRC Press
4 Park Square, Milton Park, Abingdon, Oxon, OX14 4RN

CRC Press is an imprint of Taylor & Francis Group, LLC

© 2024 Sharon Ann Holgate

ISBN: 978-1-032-07422-1 (hbk)
ISBN: 978-1-032-06911-1 (pbk)
ISBN: 978-1-003-20682-8 (ebk)

DOI: 10.1201/9781003206828

Typeset in Garamond
by codeMantra

For my dear friend Rob,

for still being here for me after all these years

Contents

Preface

I first had the idea for this book in the summer of 2021, shortly after the UK had emerged from its third lockdown of the COVID-19 pandemic. Communication in various forms had been on my mind—ranging from the health messages being given out, to the moves many of us had needed to make to online forms of contact for both work and social purposes. I had also been considering how many people had contacted me over the past few years asking for advice on science communication.

My train of thought was no doubt equally being triggered by the fact that I was approaching the anniversary of my first 25 years as a science writer and broadcaster. This mid-career milestone caused me to reflect on how valuable all the support and advice was that I had been given by various mentors and more senior colleagues over the years. When in 2022, I was awarded the prestigious William Thomson, Lord Kelvin Medal and Prize from the Institute of Physics for public engagement in physics these feelings were compounded.

I felt a stronger than ever desire to pass on to others a lot of the knowledge about science communication that I had amassed, not least because many of my mentors had by this point sadly passed away. Since 2019 I had also been enjoying my teaching duties as a guest educator on the Science Communications module for fourth-year MSci and MSc students in the Department of Physics at King's College London. All in all, it seemed a timely moment to think about how I could best collate and share a lot of the skills and advice I had been accruing to help people who might not have similar access to this information.

For this book to be as useful as possible for a wide range of scientists at different career stages, I have split it into concise chapters. My aim is to take readers straight to the specific information they need as speedily as possible when a particular communication situation arises. This allows for quick reference for anyone new to a particular communication method, or an instant refresher.

However this does not preclude reading the book from start to finish. In fact, I have opened with a selection of chapters containing information that is general to many different forms of communication. That way, if readers have the time to study the book from the start, they can sequentially build up skills, and gain a fuller picture of communicating science. This approach makes the book equally suitable for university courses in science communication, or for self-study as part of continuing professional development.

Of course, I cannot cover every communication scenario that could be encountered. But in this book, I've aimed to provide advice on the issues and topics that students and scientists most frequently ask me about.

As with any of my books, I could not have created this entirely by myself. I am indebted in no small way to my editor (Commissioning Editor for Physics) Rebecca Hodges-Davies at Taylor and Francis for numerous helpful discussions about the concept and contents for this book, to Senior Editorial Assistant for Physics Danny Kielty for assistance with various queries, and to Production Editor Todd Perry for help and support throughout the production process. Equally Production Editing Manager Randy Burling has been invaluable for assistance with various production aspects, including helping me choose what I hope will prove to be an easy-to-use layout and format for the book. My thanks are also due to my numerous friends, colleagues and other expert interviewees who kindly gave me their thoughts on communication. They include Kathryn Adamson, Alexia Alexander-Wright, Gary Bates, Owen Bennett-Jones, Dawson Chance, Makaiko Chithambo, Laban Coblentz, David Culpeck, Francesca ("Frankie") Doddato, David Faux, Roger Fenn, Andrew Fisher, Mel Harvey, Richard Keegan, Peter Main, Julian Mayers, Mario Merola, Paul Parsons, Beth Price, and Rob Scovell. In addition, friends and family—including my mother Joan, and Emma and Chantelle Winder—have provided constant support and a sounding board for ideas.

Sharon Ann Holgate
Sussex, U.K., March 2023

Author

Sharon Ann Holgate has a doctorate in experimental physics from the University of Sussex in the UK, where she was a Visiting Fellow in Physics and Astronomy for 9 years, and is a Chartered Scientist and Chartered Physicist. She has worked for 25 years as a freelance science writer and broadcaster, with broadcast credits including presenting on BBC Radio 4, and on a Saturday morning youth programme on the BBC World Service with four and a half million listeners, presenting video podcasts for medical research charity the Myrovlytis Trust, and appearing on a 'Boffins Special' of The Weakest Link. Her articles have appeared in *Science, Science Careers, New Scientist, The Times Higher Education Supplement, The Times Literary Supplement, E&T, Flipside, Focus, Physics World, Laser Systems Europe, Interactions, Materials World, Modern Astronomer,* and *Astronomy Now,* while her first book *The Way Science Works* (a children's popular science book co-authored with Robin Kerrod) was shortlisted for the Aventis Prizes for Science Books Junior Prize. Her latest popular science book *Nuclear Fusion: The Race to Build a Mini-Sun on Earth* was published in 2022. She was a contributor to the popular science books *30-Second Quantum Theory* and *30-Second Energy,* and her undergraduate textbook *Understanding Solid State Physics* is currently in its second edition and in use as a core text in universities around the world. Sharon Ann has also written three books for her *Outside the Research Lab* textbook series; *Volume 1: Physics in the Arts, Architecture and Design, Volume 2: Physics in Vintage and Modern Transport,* and *Volume 3: Physics in Sport.* In addition, she has written careers material, case studies, and press releases for the Institute of Physics, and careers material and brochures for The Institute of Physics and Engineering in Medicine, and given talks at venues including the Science Museum in London. Sharon Ann is the 2022 recipient of the William Thomson, Lord Kelvin Medal and Prize from the Institute of Physics for public engagement in physics, was the Institute of Physics Young Professional Physicist of the Year for 2006, won a Merit Award in the 1994 Daily Telegraph Young Science Writer of the Year competition, and was shortlisted for the radio programme category of the Association of British Science Writers' Awards in 2005. Outside of work she collects contemporary ceramics, is a regular visitor to art galleries and museums, enjoys learning about fashion history and steam locomotives, and keeps an ever-increasing variety of cacti and succulents. You can follow Sharon Ann on Instagram @everydaysciencethings

(Author photo: Stuart Robinson, 2022. © Sharon Ann Holgate).

1

Introduction

First Steps

For many of us, the first experience we have of communicating science outside of college or university is at a scientific conference. This is a challenging task for most people, but fortunately there is so much more to science communication than giving a good conference presentation. There are a wide range of scenarios in which you may be required to explain science clearly, and the types of communication you need to do are likely to change as you progress through your career.

They could also change much more rapidly. For example, at the start of a week you could be discussing the results of your recent experiments with a research collaborator, and by the weekend you might be engaging with the public at your lab's open day. The idea of this book is to help you through many common communication situations like these, no matter what scientific career path you take.

If you work in industry or academia for instance, you will probably need to give talks and presentations to both small and large audiences (Figure 1.1), possibly at major events.

DOI: 10.1201/9781003206828-1

1

Plus in many scientific job roles, there is an expectation that you will take part in science outreach activities for the general public or schoolchildren. This could be via talks and events, written articles, broadcasting, or social media. Equally, in just about every workplace scenario, there will be a need to explain either verbally or in writing what you are working on to colleagues (Figures 1.2 and 1.3) including line managers, people you are supervising, project collaborators, and interview panels.

So science communication will be an integral part of your future career, and could encompass a variety of media. Furthermore, good science communication skills are just as important whether you are addressing millions of people or just one.

Science communication will be an integral part of your future career

FIGURE 1.2
Chapter 11 gives advice on explaining science to supervisors and managers, while Chapter 3 will help you to tailor your communication of that science so that it is easily understood. (Shutterstock ID: 1667850958.)

1

We Need to Talk

Make no mistake about this, science communication is not some sort of 'easy option' to study nor a skillset that everyone automatically has. Get a public health message wrong, for example, and people could potentially die as a result. Make a real mess-up of a conference talk and stall your research career. Write a rushed or ill-conceived Tweet or Instagram post concerning some joint research and risk destroying a good working relationship with a collaborator. I could go on and on, and, believe me, I've heard plenty of cautionary tales over the past 25 years in this job.

But I don't want to scaremonger. With a decent amount of preparation and a methodical approach, there is no need for any of these doomsday scenarios to occur. The advice contained in this book will help you learn best practice and pick up a range of tips to improve your science communication.

1

FIGURE 1.3
Chapter 6 will help you prepare for talks of any size, while Chapter 11 gives advice on explaining science to colleagues. (Shutterstock ID: 1921582262.)

Of course I can't possibly pre-empt everything you might face. But my aim is to help you develop your communication skills in a 'practice' environment without the real-world fears of potentially losing your job/destroying a research collaboration/harming someone's health or wellbeing, or any other of the very serious consequences that can happen if you get science communication wrong. That way, no matter what science communication challenges you encounter in your future education or workplace, you will not feel thrown in at the deep end.

It is also important to understand what this book is not. It isn't meant to be professional training for would-be print, radio, or TV journalists, budding filmmakers, or aspiring social media influencers. This book is intended for people who have to, or choose to, communicate science as part of their work.

An Ever-Changing Situation

When I first became a science writer and broadcaster in the late 1990s, science communication with people outside of science was in a very different place. If I covered overlaps between art and science for newspaper and magazine articles for instance, I noticed quite a deep sense of mistrust between the two disciplines. As I moved into broadcasting, I became more and more aware that some people considered science to be something separate from their lives, and something that is really tricky to understand.

Needless to say science can be tricky to understand! I have interviewed scores of scientists over the last quarter of a century, and the challenge of trying to make sense of complex situations, or seek solutions for seemingly unsolvable problems, drives many of them. For most scientists just because a topic is complex, or difficult to grasp, does not mean they give up trying to understand it. On the contrary, it is often the most challenging issues that are the very thing that science is dealing with on a daily basis. Similarly, if scientific concepts are difficult to explain to someone who is not working in a similar field, it does not follow that we should simply give up trying to describe those concepts. Scientists don't quit when a scientific problem is stretching their abilities. So why should any of us not even attempt to talk or write about research in a way that non-experts can understand?

When we give up trying, it leads to a feeling of isolation for the person who cannot grasp the concept and the impression that science is only for an elite. In reality nothing could be further from the truth! Much of the science we rely on in our everyday lives, often without giving it a second thought, has been publicly funded by taxpayers. It could therefore be strongly argued that scientists have a duty to be accountable to the public who are effectively paying their wages. In fact, many academic and research institutions have long recognised this, and increasingly communication with the public is becoming an expected part of the job for many scientists. Science is also employed within a range of other disciplines from medicine to sport to art restoration. So with cross-disciplinary collaborations or working commonplace, it is vital for scientists to develop the skillsets required to communicate well with experts from other fields.

> Communication with the public is becoming an expected part of the job for many scientists

Horses for Courses

While my first few chapters cover some core fundamentals for all science communication, there are various communication scenarios that require

1

specific skillsets and know-how. In this book, I aim to give you these too. Armed with the right knowledge, communication tasks will seem less daunting, and should even become an enjoyable part of your career—at whatever stage you are in your working life.

You can work your way through the chapters in sequential order. Or if you need advice on a specific aspect immediately, I have written each chapter such that it can stand alone if required. That said, there is some overlap which is unavoidable. For instance, if you are giving a talk remotely you may wish to read both Chapter 5 on using remote technologies as well as Chapters 6 and 7 on preparing for and giving talks respectively. Chapter 2 on how to keep your messaging concise and simple, Chapter 3 on tailoring to your target audience, and Chapter 4 on personal presentation are also likely to be useful.

Clearly some skills are interchangeable. For example, many of the tips for speaking on the radio, which are introduced in Chapter 8, will be just as valid for podcasts, which are also explored in that chapter. But equally I have separated out different media and audiences(where it makes sense to do so) in order for the distinctions between them to be clearly seen. As we will learn in Chapter 3, one of the most important aspects of science communication is to be able to identify your target audience as accurately as possible. And understanding the strengths and limitations of different media is a crucial part of being able to successfully tailor content to your audience.

Certainly, the types of media you might need or want to use could change over time, not least as the result of new technologies or media being developed. We've certainly seen a big change in the past couple of decades:

> Understanding the strengths and limitations of different media is a crucial part of being able to successfully tailor content to your audience

when once you needed a publisher to print your writing and disseminate it to the general public, with the rapid growth of social media (and self-publishing) you can now go for it alone. But while sharing scientific information online has a unique set of pitfalls (some of which are highlighted in Chapter 10), it still has a lot of common ground with other forms of science communication. So I am confident that the basics of good communication will remain valid no matter what media come into use in future years.

From Thermoluminescence to Time Travel

Before moving into the tips and advice, I thought it might help if I first explain how I made the transition into becoming a freelance science writer and broadcaster, and what some of the challenges were. As you will no doubt have guessed from the title of this subsection, my doctoral research area was thermoluminescence, and I took both my undergraduate physics degree and doctorate at the University of Sussex in the UK.

It was actually during my DPhil that my journey into science writing began. I had just left a research seminar at Sussex when I saw a poster on the corridor wall for the Young Science Writer of the Year competition, which was at that time run by *The Daily Telegraph* national newspaper. I had already been considering becoming a professional science writer, so I duly wrote my first piece and entered it into the competition. To my delight, I made it through to the final round and ended up with a Merit Award.

Having got that far in the competition, I thought: 'well, this is obviously something that I am capable of doing'. So I wrote a handful of articles while I was a student, then after graduation began sending my CV and samples of my work off to various magazines and national newspapers—some of whom I subsequently began writing for.

The broadcasting side of my career also began when I was a doctoral student when I volunteered at my local hospital radio service, presenting some of the afternoon shows for patients. From there I progressed onto presenting at a larger charity radio station, appearing as a regular guest on BBC local radio, and eventually presenting two mini-series on the BBC World Service on their Saturday morning youth programme *The Edge* which had a regular audience of 4.5 million listeners. Since then, I have gone on to write and present my own programmes for BBC Radio 4.

Along the way, I have appeared as a guest on TV and radio shows—including being a contestant on a Boffins Special of *The Weakest Link*, presented online videos for medical research charity The Myrovlytis Trust and to accompany my textbook *Understanding Solid State Physics*, and written for a diverse range of projects including case studies for policy makers and national careers leaflets for scientific institutions. I have covered a wide range of topics in my writing and broadcasting from new cancer therapies and breakthroughs in electronics, to time travel and the science and history of toilet paper, and given talks at venues including the Science Museum in London. Recently, I have been interviewed about nuclear fusion—which was the subject of my first solo popular science book—on BBC Radio 4's *The Curious Cases of Rutherford and Fry*, on *The Bunker* podcast, and on *The Future Of... with Owen Bennett-Jones* podcast.

This wide range of commissions has meant interviewing an equally broad spectrum of scientists over the past quarter of a century. So I have sat both

1

sides of the fence—appearing on shows and being quoted in the print press, as well as conducting both print and broadcast interviews of scientists. Being an interviewer or interviewee throws up a gamut of challenges, and there is advice on the latter in Chapter 8 based on both my experiences and those of other experts.

When making the transition from a research lab into the media one of the things I found the most difficult to adjust to was not using incredible levels of detail. The fact is that whilst intricate details are essential for science, they are generally neither required nor desirable when writing or broadcasting for public audiences. But after years of working in an environment where measurements down to the nanoscale were of the upmost important, this was a hard transition for me to make and is something I will discuss further in Chapter 9.

Learning how to cope with things going awry, including some of the emergency situations covered in Chapter 13 and incidents during job interviews (which are discussed in Chapter 12), was an equally tricky and steep learning curve.

A Range of Expertise

Although as you will have just heard, I have personal experience of a wide range of communication situations, it would be impossible for me to have encountered every scenario in which science communication takes place. I am also aware that my thinking on these issues may not be the same as that of my colleagues. So in order to provide you with as broad a range of experience and views as possible, this book contains advice from several science professionals working in different sectors as well as some entertainment professionals.

The entertainment professionals offer a unique insight into communicating with others as well as some incredibly useful transferrable tips for interacting with audiences. Equally illuminating is the advice from scientists who have had first-hand experience of being on the receiving end of both good and bad communication.

So come with me and let's begin our journey into the complex, challenging, and fulfilling world of science communication.

Keeping It Simple and on Message

A Matter of Fact

"You drove me round the bend double checking everything," said one of my favourite editors, sipping at their drink. I was mortified. While I struggled to maintain a neutral expression, they continued to look me in the eye. "Best writer I ever had," they added. Phew!

I can certainly see why my being such a stickler for accuracy could be maddening in a busy publication where everyone is chasing incredibly tight deadlines. But getting the facts right is, of course, an essential part of journalism, and this editor was no exception in requiring the highest standards.

Obviously accuracy is vital for all communication of science whether that be with our peers or superiors in a work setting, or with the wider public.

DOI: 10.1201/9781003206828-2

It goes hand in hand with the need for simplicity, and the ability to tailor the level of the science to the audience you are trying to reach. In this chapter, we will look at some do's and don'ts for communicating with the public, learn how to use pictures and illustrations most effectively to help explain science, and see a variety of tips for keeping your explanations concise and on message.

It's Perfectly Logical

2

I'd say that one of the most important things when communicating science is to make sure you discuss the topic in a logical and compelling way—a bit like telling a story. To give a good starting point, in Box 2.1 I've listed the basic questions that always need covering if you are talking or writing about scientific research.

Box 2.1   Essential Content

- Who?
- What?
- Where?
- When?
- Why?
- How?

Some of these things are obvious. Clearly you need to explain what the work is, when and where it was done, and by whom. But the 'why' can be less straightforward. It encompasses things like the potential impact of the work, and why scientists chose this route of study in particular. Sometimes this can be for very personal reasons, such as trying to find a solution to a loved one's health issue. In cases like this, it is a must to write or talk about the work in a way that is sensitive to the situation.

Although I've labelled the points in Box 2.1 as 'essential' content, depending on your brief and target audience you may not in fact choose to feature all the possible answers to these questions in an article, talk or broadcast interview. You may also opt to only partially answer questions such as how the work is being done. This is because if you are writing for, or addressing, the general public your audience is not very likely to be interested in minute details such as the exact tolerances of a particular screw holding a piece of equipment together. Well, not unless the entire experiment you are discussing hinges on that! Instead you are best aiming for a broad summary of how the results were obtained, which in most cases is all most audience members want to hear or read. (Chapter 3 contains more advice on tailoring content to your audience.)

Things to Avoid and Things to Embrace

Further to this, I have a few 'don'ts' that are general to pretty much all science communication. One of these is to steer clear of using jargon. That said, sometimes you have to use a scientific term that your audience may not be familiar with. So in this case I would explain the meaning carefully the first time the term is used, and possibly even recap that meaning later on if the article or talk is quite long and there is a danger the readers or listeners might have forgotten by then.

2

Another 'don't' is using long, convoluted sentences. When I'm writing, if I'm worried a sentence is becoming too long, I read it

Steer clear of using jargon

out loud. If I can't get through it without taking a breath, then that is my cue that I need to either split the sentence up or add more punctuation!

To make sure that we're not just focusing on negatives, I also want to talk about some do's. For written science communication, even if you're writing for a professional publication where it is likely to be replaced, try to come up with an arresting title and stand first (the stand first is the sentence before the main article starts, often indicated by a different font). In some cases, these will be what attracts your readers to the piece, and at the very least they help the editor to understand the overall slant of what you've written.

With any type of science communication what you're looking to do is keep your readers or listeners engaged by holding their interest, so you need to think about how what you are writing or saying flows. In articles, one way of keeping people on-board is to use a sentence at the end of a paragraph that deliberately leads the reader on to the content of the next paragraph. You can sometimes even set up a cliff-hanger, as you can with audio. In that case, your reader or listener will be desperate to stay with you to find out what incredible result stopped the scientist in their tracks!

Obviously, you can't use devices like this constantly without it sounding a bit peculiar. But you must aim to write or deliver content in such a way that the audience is hooked and eager to find out what happened next.

Having a narrative thread running through so that you are telling a story is another way of achieving this.

It's Like That, and That's the Way It Is

One thing that works really well for explaining science to non-scientific audiences, and especially children, is analogy. So you will often see things described in terms of size in relation to a football pitch, or perhaps a familiar everyday object (Figure 2.1). As I revealed in Chapter 1, this was one of the adjustments I actually found the hardest to make when moving from research science to science writing and broadcasting: coming from a laboratory environment where a very high level of accuracy was used constantly then suddenly having to switch to comparing objects only vaguely on the same scale. It's a totally different way of thinking, but you have to realise that most people can't recite pi to goodness knows how many digits. That sort of accuracy has no relevance to a lot of people's lives.

FIGURE 2.1
Relating sizes to familiar everyday things, from fruit to football pitches, can help audiences to visualise scientific topics. (Photo: © Sharon Ann Holgate.)

Another good way of describing scale, that everyone will be familiar with, is to compare the size to parts of the body. For example, comparing a particular width to that of a single human hair is instantly relatable. (Chapter 3 contains more advice on using analogies in your communication.)

Summarising

To help ensure the audience gets a 'take home' from any talks that you give, it is a good idea to recap key messages at the end of your presentation. This will allow for the fact that people may have drifted off, or might not have fully grasped the main points you were trying to make. The latter is a particular danger if you have been going into a lot of detail about methodologies or results.

With this though, always bear in mind that what might seem to be obvious to you may not be obvious to your audience. I

> Recap key messages at the end of your presentation

once sat through a 20-minute YouTube video on how to run a large physics experiment. The video told me everything except for exactly what was happening in each experimental run. This was the only piece of information I needed, and I didn't get it! So as well as getting your own key messages across it is worth thinking about what readers and listeners might most want to know, and including this information as part of your summary.

Making It All Add Up

One of the most challenging aspects of science to talk or write about can be the numbers or mathematics associated with it.

"Maths is a very, very tricky thing to write about," says Dr Roger Fenn, an Emeritus Reader in Mathematics from the University of Sussex in the UK, who has written articles for popular science magazines. He stresses that the public's mathematical understanding can vary widely, so you can't get away with assuming prior knowledge in mathematics.

"Some people get to fractions and then give up. Some people go past that [stage] to being able to draw a graph of a function, and then some people go beyond that," he explains. To cover all bases when writing or talking about maths for a general audience, Roger advises that "the main thing to do is to use stories rather than jumping straight into theories. Bring up personalities from maths or science, and explain what they thought or think about what you are describing."

2

FIGURE 2.2
Using diagrams can help explain the mathematics connected to science. (Shutterstock ID: 1880747335.)

He also favours using diagrams or pictures to help explain concepts. "In physics you can compare things like the size of an atom with the size of the solar system. In maths these sorts of comparisons are trickier" (Figure 2.2).

Even seemingly simple numbers can prove challenging to describe. "People talk gaily about billions and millions, and actually there's a huge difference [between them]. I think the best way of showing the difference is either with a line so you show a million as a tiny bit of your line and a billion as much further along the line," suggests Roger, adding that different sized areas can similarly be represented by different sized or coloured dots on a diagram.

He warns that not everyone fully understands the most commonly used mathematical terms and concepts either. "Politicians often misunderstand inflation, for example. They think that if they get inflation down prices will go down. But of course they won't. Unless inflation goes negative, you're not going to get prices down because inflation just means the difference between two prices—it's not the absolute value of something. So that is a total misconception," he continues, stressing that it is important to be really clear about what maths and numbers actually mean when you are writing or talking about them.

From Bottom to Top

Needless to say, communicating with the public in simple terms is about much more than just employing some good analogies and reducing the level of technical difficulty compared with speaking to scientific colleagues. You need to completely change the order in which you present scientific information, according to Laban Coblentz, Head of Communication at the ITER Organization in France, the world's largest fusion project.

"If you look at a scientific paper, you might have an abstract that you can read which is relatively short, but otherwise the flow of information is presented by: fact, experiment, gathering data, and a long accumulation of information that leads you downward to a well-substantiated conclusion. So your conclusion comes at the end. If you use that model in journalistic communication, you'll never get anywhere because the model for journalistic communication is exactly the reverse. You need to have in the title, or at least in the opening paragraph, something that states your conclusion or end point so clearly that it will draw the reader on to read the rest of the story," explains Laban, who studied communications, English, and psychology at University prior to studying nuclear engineering and reactor physics while in the US Navy, and subsequently working in a range of science communications roles.

While hooking the reader or listener in has always been necessary, nowadays doing this extremely rapidly is more important than ever, feels Laban, who has trained engineers and scientists all around the world in how to speak to a lay audience.

"About 10 or 15 years ago we passed from an age of information into an age of information overload. The result is that while we used to be pleased that we were getting 10 or 15 emails a day from trusted colleagues or our family, what you're seeing now is a completely different information universe. I am now getting 200–250 emails a day, which means that I make a decision within a matter of 2 or 3 seconds as to whether it is something I have to read, something I can delete immediately, or something I can set aside for further perusal at a later date. That emphasises the need to be very skilled in what you put up front."

These up-front sentences or statements will need to be extremely concise. But the good news is that being succinct is an ability most of us have already honed. "If we

> About 10 or 15 years ago, we passed from an age of information into an age of information overload

are sending a text to somebody, we all know how to compress that information into something short. If I'm on a news programme where the merit of my interview is whether as a science communicator I can compress something

2

FIGURE 2.3
Sometimes you only get a few seconds to make your
point on radio or TV, so ensure you are concise.
(Shutterstock ID: 1612772830.)

into 5 seconds or 30 seconds, it is the same skill," says Laban, acknowledging
that "we are increasingly given a shorter and shorter window with which to
make our impression and our conclusion" (Figure 2.3).

"If you adhere to the classic format of a scientific paper in your communication with the public, by the time you reach the conclusion the audience
is either asleep or mired in detail beyond their comprehension," continues
Laban, whose past roles have included ghost-writing speeches for prominent
scientists to deliver. Getting to a rapid conclusion or a soundbite is especially
important for live broadcasts where time is limited, he feels. Otherwise the
risk is that you only deliver part of your message. Speaking in short sentences is also important. "Otherwise the audience can have completely lost
the thread of what you are saying by the time you have reached your fourth
or fifth clause," he says. (Chapters 6 and 8 give more advice on dealing with
media interviews.)

A Picture Tells a Thousand Words

To simplify messages when giving talks, Laban suggests replacing any of your PowerPoint slides that are full of words. "Use the same concepts, but look at breaking it down into either no words or very few words, and use the force of an illustration with a shorter version of the text." You can then talk about the topic concisely rather than having everything written up on the slide, he says. "So move to more visuals which tend to be more remembered, and think about the relative complexity of your explanations."

As long as you don't allow seeing your-self on camera to put you off, recording your entire presentation and playing it back can transform your presenting, and make sure you are keeping it simple, feels Laban. "It's a way for you to see if you need to look up more, or change your body language, or speak in more simple phrases," he explains. (Chapter 6 contains more tips on preparing for talks.)

Use the force of an illustration with a shorter version of the text.

2

Being Open

It should go without saying that keeping on message should never be to the detriment of telling the truth. But this can be an easy trap to fall into if you are not very careful.

Taking the difference between nuclear fusion and nuclear fission as an example, Laban Coblentz's personal view is that it is unhelpful to illustrate how much safer fusion is compared to fission by calling it something like 'fusion energy' or 'hydrogen fusion'. By refusing to associate the word 'nuclear' with the process, he feels that risks "a severe undermining of trust" from the public when other experts then reveal that fusion is in fact a nuclear reaction. So he advocates communicating openly and not trying to cover up certain aspects of science, no matter how well intentioned your reasoning behind that decision.

Coupled with this, Laban feels we should not underestimate the public's ability to grasp nuanced scientific concepts. "We do ourselves a disservice by perpetuating a legacy that members of the public believe they cannot understand complex science. The norm when you're trying to explain something is to use phrases such as: 'don't worry that's not rocket science', or 'don't worry that's not nuclear physics'. It's 'not that tough to understand' is what we mean, and conversely that it's impossible for a 'normal' person to understand rocket science and nuclear physics. And it's not. It's not difficult to explain most of these concepts whatsoever."

"I refuse to acknowledge this barrier. I regard our public scepticism about science as a legacy problem that we will eventually overcome," continues Laban. "The public are not incapable of understanding advanced technology, and many things can be explained very simply," he adds, pointing out that he can show his 80-year-old Aunt how to swipe open a smartphone without her needing to understand all the binary numbers behind that command operation.

> The public are not incapable of understanding advanced technology

2

Tailoring to Your Target Audience

Tailor Made

"Blimey, you're really scary!" exclaimed the radio producer I was working with. This was after I'd just replied "you could say that, but it would be wrong" in response to their asking if a sentence they'd written summarised the science OK. I learnt a great deal from working with this producer, and really enjoyed the experience as I got on so well with them. This particular incident was certainly an important lesson for me. It illustrated not only how differently people without a science background process information compared with folk who have studied or worked in science, but also how much I would need to alter my explanations to account for that.

DOI: 10.1201/9781003206828-3

There are many different audience demographics that you may need to communicate with throughout your career, and each one of those will be unique to some extent. However, in this chapter we will see tips for communicating with some of the audience types you are most likely to encounter—namely fellow scientists, family audiences, and the general public. This chapter also includes valuable information on addressing overseas audiences, and on ensuring your content is accessible for disabled audience members.

Getting Into the Right Mindset

But before looking at some examples of tailoring to specific audiences, it is worth considering how to approach the preparation of your content for any given audience. You need to think carefully about this, especially if you have not addressed that audience demographic previously.

3

So what are my recommendations for getting into the right mindset for serving your target audience?

First, have a think about whether you have had first-hand experience of being in that demographic yourself. For example, we were all children once and it can help to think back to what sorts of science things you used to enjoy seeing and hearing about. Equally, recalling what you disliked as a child when it comes to science is helpful. (I, for instance, could not stand 'whizz-bang' type experiments, but would sit glued to TV programmes about space travel even when I was too young to fully comprehend them.)

Similarly, recalling your preferences at the various stages of your student and/or career journey is valuable. You can even take this a step further and analyse how your preferences have changed over time. Did dramatic experimental demonstrations first get you interested in science, or did you prefer to sit quietly reading books when you were younger? What speed of delivery for scientific content did you like best as an undergraduate student compared with how you feel about this as a postdoc? (Figure 3.1).

> Think back to what sorts of science things you used to enjoy seeing and hearing about… and analyse how your preferences have changed over time

Any comparisons like this are helping you to step into the shoes of your audience. You can even imagine being a non-scientist by considering how you react to scientific information from fields far outside your own that you only have a very limited knowledge of.

If you read or watch enough content, you should soon start to get a feel for the right sort of approach. But if you are still uncertain, try speaking to someone in that demographic and get their views. This is helpful not least because it will give you an insight into other people's tastes and preferences.

FIGURE 3.1

Think through the ways in which you like to learn have changed over the years to help get you into the right mindset for communicating to specific age demographics. (Shutterstock ID: 150780812.)

3

As well as thinking about the speed and type of delivery that should be suitable for your target audience, you also need to ask yourself what level of prior scientific knowledge they are likely to have?

This can be easier said than done if you are addressing the general public. Your audience could contain a research chemist from a pharmaceutical company alongside a solicitor who last studied science at school. There is no way of knowing. To cater for this unknown, I generally try to include interesting details that practising scientists may not know, but make sure they are described in a way that people with a very limited level of scientific knowledge would understand. No one wants to hear masses of complex technical details, no matter what their background. (See Chapter 9 for more advice on keeping to the main facts.)

But no matter how much you consider such aspects, there is no way of pleasing every single audience member. Also, you do need to try and find your own voice. In fact, you'll notice that the tips given throughout this book are not instructions for completely curbing your style, and adopting some sort of 'standardised' way of writing and presenting. The last thing I would want to do is attempt to stifle individuality.

That said when it comes to writing or broadcasting if you are commissioned by an outlet, rather than simply pasting your own unedited content online, you are likely to be subject to a number of constraints. These could take the form of a house style, or a specific length or slant. But even then I still feel that it's really important to tread your own path and not just copy other people's approaches.

Age Appropriate

I certainly needed to come up with some innovative approaches to science communication when I teamed up with professional magician Tony Drewitt a few years ago to present some science and magic shows at the Jewish Museum in London. These shows were for family audiences, so there were a lot of kids there, and we used illusions to help explain scientific topics.

One of the things we covered was the astonishing fact that an adult locust can eat its own body weight in food daily. Now you might think yes, well, it's a relatively small creature so it's not eating that much. But if you scale this idea up, it gives a much better feel for what an incredible scientific fact this is. So I came up with a menu of what I would need to eat in order to consume my own body weight in a day, shown here in Figure 3.2. (I should add by way of a disclaimer that I don't recommend following this menu!)

My 'menu' concept went down well with our audiences. But something that definitely does not work well for any audience is using language that could be construed as rude or vulgar—even if you do so in jest. It is especially offensive if there are children in the audience, but I'd steer clear of that for any demographic. I've occasionally had words added in to articles by editors that I certainly wouldn't have chosen to write, and there's nothing I can do about that. But in my view it's always best to leave expletives and 'rude' words out because you don't know whether you might offend your audience. If you do, then they could switch off from the science content, thereby defeating the whole object of the exercise.

I'd also suggest caution when using slang. The main problem is that slang is often specific to a particular age group. For example describing something as 'sick' will have a totally different meaning to a 15-year old and a 70-year old. There may be instances where an audience has a very specific demographic and you can use age tailored language. But you do need to be

BODY WEIGHT
RESTAURANT

FREE DELIVERY
+123-456-7890

LUNCH

Large pizzas	20
Bowls of salad	30
Apples	75

BREAKFAST

Cornflake Boxes	10
Bags of sugar	4
Grapefruit	20
Loaves of Bread	7

SUPPER

Double cheeseburgers	35
Tubs of ice cream	10

3

absolutely certain that you know what the slang means, particularly if you are not in that age group yourself (Figure 3.3).

Slang is often specific to a particular age group

Pitching it Right

When explaining science to fellow scientists, although you won't have children in your audience avoiding overloading your content with details, and choosing relatable analogies is equally useful.

"If I give a speech at a fusion conference, I know that the audience understands fusion. So I don't need to explain the basic principles. But to keep the attention of the audience, you equally cannot go into too much detail. Most of the people will just need to know what the main difficulties and challenges are, and how have you resolved them," says Mario Merola, Head of the Engineering Design Department at the ITER Organization in France, the world's largest fusion project.

3

Mario regularly delivers talks internationally to both specialists and the general public, and he recommends the use of analogies when speaking to all types of audience. "One important thing that I know is effective is to make a comparison with something which is tangible in the daily life of the audience. So when I explain the heat flux coming from the plasma onto the plasma facing components I make a comparison with the heat that you feel while you are lying down in the sun in the summer. Also, when I have to give a speech on the overall ITER project, then I show the vacuum chamber where the plasma is contained alongside a picture of the Eiffel Tower in Paris. I say that the mass of the Eiffel Tower is the same as the mass of the vacuum chamber in ITER. So it is something that people can better appreciate and compare with things in their daily life," he says, adding that he finds analogies like this just as effective when he is presenting at scientific conferences as when he engages with the public.

> Make a comparison with something which is tangible

Overseas Audiences

Of course events can be anywhere in the world, and for many scientists speaking at international conferences and overseas meetings is an integral, and regularly occuring, part of their career. It goes without saying that giving a presentation overseas requires additional planning.

Preparing for events in other countries not only involves a lot in terms of sorting out travel arrangements. You will also need to think carefully about the content of your talk to ensure that any context and caveats will be understood by your audience.

So what are the main dos and don'ts for talking to an overseas audience, especially when you are speaking in a language that is not native to many in the audience? According to Prof Makaiko Chithambo, Head of the Department of Physics and Electronics at Rhodes University in South Africa, you need to ensure that you "don't change who you are. Don't change your voice. Just speak the way you naturally do."

Speak the way you naturally do

3

He does however caution against talking about things in exactly the same way that you would in your home country. "Avoid colloquialisms that would be understood locally, but not necessarily internationally," says Makaiko, adding that if you are forced to explain what you are saying, this can result in subtleties in your meaning becoming lost.

Another thing to keep a check on is your speed of delivery. "Don't deliberately slow down," cautions Makaiko. "It's OK to slow down for a bit to make a point. But if you make your whole talk drawn out and dragging, it almost feels as if you are talking down to your audience. So I would say: avoid that."

Knowing the scientific level to pitch your talk at will assist with your delivery, he feels. "If you understand your audience it helps you come across naturally. A group of students or a group of academics are likely to understand your jargon and technical details," explains Makaiko, adding that academic audiences will generally have been exposed to your local style of diction from university lectures, YouTube, or TV.

Even so, Makaiko admits that he gets ready to listen hard if he knows a speaker comes from a part of the world from which he tends to find certain pronunciations difficult to tune in to. "You inwardly prepare yourself to understand because you know for that speaker English will be their second language as well," he says.

"From my own [presenting] experience, I would say to give hints to your audience about what you're saying. So for instance, if I am making a PowerPoint presentation I might read a sentence from a slide. Then if a member of the audience is having trouble understanding, by hearing me say certain words they catch on to my pronunciation and so what it is I'm talking about. Because unless your diction is very clear it does take time for most people to understand," continues Makaiko, who feels it can take around five minutes to fully tune your ear to the accent of a speaker.

He says he takes extra care when speaking in English to make sure his audience knows how he is pronouncing words that do not sound how they appear when written. He also advises double checking beforehand how to pronounce people's names and place names. "Be careful as some people can be sensitive when you mispronounce things," warns Makaiko.

Accessible for All

3 No matter which country you are speaking in, there is also another aspect of accessibility to consider and that is catering for disabled audience members.

Dr Frankie Doddato, a Teaching Assistant in the physics department of Lancaster University in the UK, disability advocate, and a Co-Chair of the Lancaster University Disabled Employees Network DEN+, explains that people could have a wide range of different requirements when it comes to accessing your communication. For example, they may have physical disabilities such as deafness or visual impairment, or they could be neurodivergent, she says. "Neurodivergence is an umbrella term for anyone who is not neurotypical. So it encompasses autism, ADHD, bipolar, dyslexia and dyscalculia [difficulty in understanding numbers], etc. Neurodiversity is where you have a group of people of different neurotypes which can include neurotypical in the group," explains Frankie, who has several pieces of advice for making any information you are imparting as accessible as possible.

"In the case of autism you want to make sure that you're very clear with your communication," says Frankie, who is autistic herself. The way autism manifests is so broad (and it's not a linear spectrum of 'mild' to 'severe'—a reason autistics can prefer to avoid referring to autism as a spectrum) that you may not be doing the optimum thing for everyone, she explains. "But you want to be unambiguous with your meaning, very clear, and use a clear tone of voice." Frankie also recommends steering clear of sarcasm and being careful with your use of comedy because people with autism might not understand the joke. "If someone is very deadpan and sarcastic that can really throw someone off what you are saying," she warns. "But that doesn't mean speaking slowly and like you would to a small child, over-simplifying things. Speak to autistics, and other neurodivergent and disabled people, like people.

A common problem is that a lot of people infantilise disabled people, which is very wrong."

If you are giving a talk, to aid accessibility you need to ensure that each slide is laid out in a clear way, and does not have too much going on because that can lead to sensory overload, advises Frankie. "Having things like lots of bright colours all over slides like a firework display can be very overwhelming," she says.

As Frankie points out, words, images, and symbols scattered over a slide could be problematic for people with visual problems to read too, as well as for neurodivergent people. Not least if there is colour on top of colour as this makes for low contrast. "So it's about finding that balance really of trying to make sure that you've got a good contrast between backgrounds and text and not too many arrows and pictures and text all over the place haphazardly, but at the same time ensuring that it is not so bright and garish that it's overwhelming," says Frankie.

"Choice of font and font size are also important. Sans Serif fonts, Helvetica, and Arial are good choices, as they have simple character shapes which can be useful for visual impairment or dyslexia, for example. For font size, 12–14pt is the best range for documents, with 14pt preferred. However, some accessibility requirements can clash. Where some autistic people might need very soft colour palettes, with low contrasts, some ADHDers can find that having pops of colour can help them to focus and pick information out. Some fonts which can be accessible for some might be less accessible for others. It's all about finding that balance."

Trying to steer clear of using writing that's faint, or presenting information on grid or lined backgrounds, is also good

practice for accessibility and clarity, she continues. In addition, "you need to make sure that any scans you use are high quality".

Frankie says that she always stresses to her students that their work needs to be laid out in a logical, easy-to-follow pattern that makes it clear to the reader what the linear progression of the working is. She feels this is just as important when preparing slides for talks. "You don't know whether your audience members are disabled or not. You may have someone who is colour blind or someone who has a sight impairment or someone who is neurodivergent. You want to make your talk as accessible as possible," says Frankie, stressing that making these best practice choices actually helps make the slides clearer for everyone in the audience. (See Box 3.1.)

Frankie also recommends using good alignment for the text on each of your slides as this can help autistic people and also help with visual processing. (Figure 3.4 gives an example of a less accessible slide.)

FIGURE 3.4
Avoid creating slides like this with text overlapping the graphic elements, grid backgrounds, and a lot of different visual components dotted all over them. (Graphic: Canva. Text © Sharon Ann Holgate.)

3

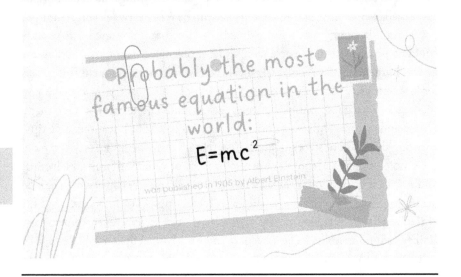

Box 3.1   Making Clear, Accessible Slides

- Use a plain background
- Have good colour contrast
- Use an accessible font such as Arial, Sans Serif, Helvetica, preferably in 12–14pt size
- Use bold font to highlight text rather than highlighting, italicising, or underlining it
- Don't cram too much information onto each slide
- Use a logical flow through your points
- Have good alignment for your text

In terms of auditory communication, Frankie says an echoing room, background music, or a lot of audience chattering can cause sensory overload for someone who is neurodivergent.

These issues can equally create challenges for people with hearing loss. So Frankie suggests employing simple methods such as making sure that the text on your slides is large enough that it can be seen from the very back of the lecture hall, and taking care not to speak too fast and to enunciate clearly. "It's about finding ways to communicate and present science that supports disability and has benefits for everyone," she concludes.

3

4

Personal Presentation

Owning it

"She shouldn't have hair that long, it's too high maintenance," said one of an elderly couple sitting near to me in a train carriage I had just boarded. They proceeded to imply that I must do something useless for a living as all I cared about was my hair! The reality was that I had been awarded a doctorate in experimental physics just a few months earlier, and was starting to write for *New Scientist*. Plus I wasn't asking them to blow dry my then waist-length hair, was I! I am presuming they thought I could not hear their commentary. But I could, and it upset me. I feel that it's no one else's business how we all choose to present ourselves, and I can't help thinking the world would be a better place if people kept their opinions on such topics to themselves. So why, you might well ask, am I writing a chapter on personal presentation? Well, the answer to that lies firmly

DOI: 10.1201/9781003206828-4

rooted in the wider context of communication and how people respond to others. Let me give an example.

Suppose you had a legal issue that you needed help from a solicitor to sort out. Would you feel confident in their ability to deal with your issue if they sat through your first meeting wearing a tropical print shirt and some beach shorts?

OK. So that's an extreme example, right? Well, not really. I've interviewed academics who have been appalled at science students turning up for PhD interviews dressed exactly like this. Clearly, alienating the very people who you are trying to have a positive communication experience with is not a good idea. So in this chapter we will look at the importance of choosing suitable clothing for those key work occasions, as well as offering some tips on personal grooming (Figure 4.1). These might just give you that extra bit of confidence when facing a big audience or an important interview, and help you avoid inadvertently hindering your career progression. Essentially, anything likely to distract either the audience or yourself when delivering a science message is what we will be focussing on avoiding.

FIGURE 4.1
It is wise to choose more formal clothing for special work events and leave your casual outfits for other occasions. (Shutterstock ID: 357764558.)

4

You Had Me at 'Hello'

But surely when it comes to science communication it is only the science you're talking about not how you look that counts, right? Wrong, I'm afraid! Like it or not, your appearance is an important part of communicating any topic, including science. This is because whoever you are speaking to will be responding to your clothing and grooming even if they don't mean to.

"First impressions are hugely important because as humans we are hard-wired to judge on appearances. It's crazy, but in the first seven seconds we make snap judgements subconsciously," explains award-winning stylist and make-up artist Beth Price, who works with clients in the corporate, education and charity sectors, and was Personal Stylist of the Year in the Cardiff & South Wales Prestige Awards 2020 and Best Personal Stylist 2020 (Wales) by LUXLife Health, Beauty & Wellness.

First impressions are hugely important

She adds that it can be very difficult to change opinions made further down the line. This makes it vital to nail your look for things like job or placement interviews. "If you come in and the interviewer instantly decides that you look competent and credible, you have to talk a lot less to persuade them you are the right person for that job. Whereas if you've made a bad first impression there's a lot more work to be done to turn the interviewer around," she says.

4

Beth points to studies by Albert Mehrabian, Professor Emeritus in psychology at the University of California, Los Angeles in the US. "In his work he says that 55% of a first impression is based on your appearance and body language, 38% is the tone, pitch and pace of your voice, and only 7% is the vocabulary and words that you are actually using," she explains.

Putting effort into your outward appearance can also help with your vocal delivery feels Beth. "If you feel like you look spot on, your tone, pitch and pace of voice will be slower and more confident. So that first impression will be much, much, better." If you have mis-judged the environment and feel really underdressed or overdressed and that you are not fitting in, "you may either put your head down and mumble, or you'll speak really quickly to get out of the situation as rapidly as possible. It all goes downhill. But when you know you look appropriate, you feel confident and you come across so much better. It makes your life easier if you can make a good first impression."

Feedback from Beth's own clients over the past 21 years has revealed that changing their look has made them feel more confident and more visible, and also that people are paying more attention to what they are saying. So planning an effective outfit and grooming strategy for any job interviews or public speaking engagements could change your life by opening up opportunities. "It's very short-sighted of the people who think you will only be judged on your experience and your qualifications because you absolutely won't'," she states.

Adding Some Colour to Your Life

So how can we best come up with a killer outfit for those important work occasions that captures people's attention and is appropriate for the setting but does not stifle our individuality?

The first thing to consider is what colour combinations to use, advises Beth, who worked for 16 years as an Image Consultant and Personal Stylist for Colour Me Beautiful, an international image consultancy for business clients.

"Colours can drain the life out of you and make you look unhealthy and tired. So you want colours that work well with your own skin tone, eye and hair colouring, and that engage your audience and tell them a bit about who you are. You wouldn't expect your accountant or solicitor to wear the same colours as a primary school teacher or an ecological scientist. They are going to care about different things, and you'll want the colours [you choose] to be saying something about you and to be memorable as well."

> You want colours that work well with your own skin tone, eye and hair colouring

This latter point is especially important for job interviews. "Interviewers see a million people in black or grey suits with a white shirt or blouse. So if you wore, for example, a chocolate brown suit and a teal blouse you'd look more individual and be so much more memorable. Or you might wear a blue shirt and tie with a pocket square and some smart cufflinks with a navy suit. (Blue is a colour that instils trust and conveys logic.) If you want to show ambition or energy you might wear a pop of red, and if you are in a green, ecological industry you might use some green in how you dress," continues Beth.

Picking a colour that is too bland can also backfire when presenting. "If you're giving a talk and you're wearing beige people could nod off. But if you are wearing a bolder colour, that will hold the audience's attention," she enthuses.

Taking to the Stage

Another thing to take into consideration when presenting is to make sure that your clothes fit properly, says Beth. She suggests avoiding things like floaty sleeves falling over your hands, and also fabrics that are either too stiff or too clingy for your body shape.

"If you are fidgeting and worrying about your clothes your confidence levels drop," she explains, adding that poorly fitting clothes can also distract your audience directly as they will be focussing on your clothes and not on you.

If you are fidgeting and worrying about your clothes your confidence levels drop

It is also important to road test clothing before an event, feels Beth. This can help avoid issues that could make you self-conscious, such as your knee-length skirt being long enough when you're standing but showing too much leg when you sit down. "Stand up, sit down, and look in the mirror and see what your outfit does," she advises.

She also recommends carefully considering your choice of underwear. A good fit will help with posture and comfort, while wearing underwear to match your skin tone will mean it won't show through your garments if there are bright lights on stage or bright flashes when having photos taken.

Fortunately for those of us on a budget, or with medical conditions that necessitate prioritising practicality and comfort when choosing clothing, a public speaking engagement does not automatically mean we have to reach for a formal suit. These days you might be on stage wearing a smart casual shirt and a pair of chinos, or a long cardigan and a soft dress. There are many options which are fine, according to Beth, as long your outfit "holds attention and shows authority."

Rather than focussing on looking very formal, "it's much more important to think about the colour and what it says, and that your clothes don't look tired and old. Sometimes it is about staying modern and fresh, because when you start to look old-fashioned people may switch off, thinking your ideas are outdated too. It's advisable to stay current and have a look at what's going on [fashion-wise], not just fish something out of the wardrobe that is smart and tidy but very dated."

Add-On Extras?

"Your accessories and your grooming all form part of that first impression," continues Beth. But while no one is going to suggest you completely stifle your individuality, in more formal industries you may need to think carefully about how much jewellery you choose to wear—especially when it comes to facial piercings— in situations like job interviews.

Your accessories and your grooming all form part of that first impression

"You are going to need to 'declutter'. For example, if you have six earrings I'd say just have one pair of neat earrings that is not going to distract the interviewer. Don't have too much going on. You don't want bracelets, five rings, multiple piercings, because then people are looking at all these things and not focussing on what you're saying. It shows good attention to detail to have just one necklace or one pair of earrings. If you are going to wear jewellery, make sure it is a bold statement and shows confidence and that you don't look like you are scared to wear it."

"Style is a way of showing who you are without you having to speak, but in more formal situations it is best to keep it very safe." The main point, says Beth, is to not switch anyone off from what you are trying to talk about. She recommends looking at what leading politicians and commentators are wearing because they have to appeal to a wide range of people and be listened to.

Your choice of clothing and accessories is very much part of your personal branding, and "you've got to be showing the values that you want to put across," says Beth. "You will always have the Mark Zuckerberg's of this world who wear their jeans and T-shirts and they become multi-millionaires despite their look," she continues, stressing that more generally a very casual approach to style does not help people get ahead. "Steve Jobs was very laid back, but even he didn't dress like that at the beginning. He dressed very much more smartly and was suited before he made his millions."

"Reflecting on both of these influential men, both are and were very aware of their personal branding. Steve Jobs moved with the times and updated his look. Mark Zuckerberg's company is all about friendliness and relatability so his laid back look attracts his fans, and his rebellious nature in formal situations gains him attention. It's his own company so his personal values can be conveyed. But when you work for an organisation, it's important to convey their values in your appearance, rather than your own," continues Beth.

Make-Up and Grooming

According to Beth, poor grooming can be interpreted as laziness, so good grooming is important in a work environment as is consistency in how you look. The latter, she explains, is vital because if you stop caring about your appearance after six months of being in a job role or of networking with the same colleagues or clients, this could be seen as a sign that you no longer care about your job or them.

I would say that the same goes for taking part in any form of science communication event—if you care about what you are doing it is important to also *look* like you care.

For anyone who chooses to wear make-up, Beth has some Do's and Don'ts. "The biggest 'do' is to make sure your foundation matches your skin tone, and to aim for a professional daytime look. You don't want to look shimmery or shiny or have a dark smoky eye—that's not appropriate for work."

When wearing a bright lipstick, she advises that it needs to match your skin tone so that it's not distracting anyone from what you are saying. "You don't want to make your look high fashion. Your make-up should just be modern and professional, and pared down rather than OTT. You want to be remembered for your amazing work and achievements not for your lash extensions or highly filled lips," she states.

You also need to wear make-up that is sustainable for you, feels Beth. "For most people, it needs to be quick and easy [to apply] so it looks like you've made an effort, but also be long-lasting. You don't want to look great in the morning but then halfway through the day it has all disappeared. So take your lipstick with you and top it up when you go to the loo."

> You want to be remembered for your amazing work and achievements not for your lash extensions or highly filled lips

As well as giving more general advice on grooming, Beth has helped clients with skin conditions who want to cover these up when presenting because it is negatively impacting on their confidence. This requires some specialist products tailored to your particular needs.

"To cover spots or rosacea, for example, you need to use a concealer followed by a colour corrector. If you have redness you use a yellow or green to conceal it," she explains, "then apply a full coverage foundation on top. You want it to look like skin, so either have a make-up lesson, or get some good advice from someone on a make-up counter." This, continues Beth, enables you to find "products that are right for you". If you wish to cover port wine birthmarks or similar, she recommends looking online for advice from the charity Changing Faces.

When it comes to hair, keeping it in good condition and well cut, and maintaining your colour if you choose to dye it are important, feels Beth. "A lot of men have facial hair these days, so make sure that is neat and tidy. Also look out for nose, ear, and neck hair and keep these trimmed," she adds.

Paying attention to your hands and nails is equally important, she stresses. "In a face-to-face situation one of the first things people will do is put out their hand to shake it. So you want to make sure yours are nice and clean and not rough."

If you wear nail polish, Beth recommends using a neutral colour unless you are particularly trying to make a statement. She feels neutrals are especially useful if you have a bit of damage to your nails or hands. "You always want to put attention to your good areas and if something is not great don't draw attention to it." As with every other part of your look, "keep things clean but understated," concludes Beth.

4

Planning Ahead for Conferences and Meetings

Whenever I have to go to a conference or meeting, I plan my outfits about a week in advance. This allows time for a trip to the dry cleaners, or for items to go through the wash. If you try to plan much earlier than this, the weather forecast is likely to be so inaccurate that it is pointless. But plan later, and you risk that star garment you can't manage without ending up in the laundry basket just when you want to wear it.

As well as making sure that the garments you want to wear are clean, check for any repairs that may be required. Holes in your clothes, loose buttons, or stray threads are never going to create a professional impression. So this way you have time to carry out the repair. (Or to dump the task on your relative/friend/significant other!)

> Holes in your clothes, loose buttons, or stray threads are never going to create a professional impression

I actually find it useful to write a list in a notebook of each day's outfit. Sometimes I also add a particular lipstick or eyeshadow colour to the list as well. This is all designed to make life much easier for me when it comes to the conference itself, as I know that I am likely to be in a rush getting ready. So I won't necessarily have time to flap around deciding on which accessories go with what, and whether my make-up will clash.

If you are attending scientific conferences and meetings regularly, it may be worth taking a leaf out of President Obama's book. During his presidency, I read with interest that he tended to wear only blue or grey suits so that he didn't need to make decisions on what he was wearing. (Disclaimer: I happen to think Barack Obama is one of the most stylish people on earth.) The beauty of a limited colour palette for your work clothes is not only that it frees up your mind for more important decisions. It also makes shopping for new items much easier, because if you stick to your chosen colour palette any fresh purchases will go with everything else you already own.

Instead of frantically pulling out scarves, ties, shirts, trousers, dresses or whatever else onto your bed to see what goes with what, imagine a world in which you simply reach into your wardrobe and anything you take out will work with that day's outfit. Bliss!

I'm aware that, this might all sound a bit boring. But we're talking about attending scientific conferences here. Not going to Coachella. (Or indeed any other music festival, for which I am reliably informed mud-proof boots are the main essential.) Limiting your work clothing options simply makes sense time-wise, and it needn't feel stifling because you can simply let rip with wearing anything you like in your free time.

4

Packing it in

Needless to say, not all conferences and meetings are close enough that you can travel daily to them, and if you need to pack I've found out the hard way what does and doesn't work. First, I would strongly recommend rolling most types of clothes rather than folding them. That is unlikely to be possible with a tailored jacket, but for most other garments I've found this leads to far less creases on arrival at my destination. I also buy a fresh pack of acid-free tissue paper, and place a layer on top of any delicate garments or fabrics that crease easily—such as linen—before rolling or folding. That way, the tissue provides a bit of a barrier between different layers of the garment (Figure 4.2).

If you arrive in your hotel room ahead of giving the biggest presentation of your life and can't figure out how to operate the trouser press, or your room does not have one, then there is no need to have a meltdown! Simply run a hot shower to get some steam into the en-suite, and hang up your creased items for a while. Obviously don't get dry clean only items too damp. But you can keep an eye on progress and watch those creases drop out!

FIGURE 4.2
Placing a layer of tissue paper on top of garments before rolling them up or folding them can help prevent your clothes from creasing. (Shutterstock ID: 2061961292.)

Likewise, if there is no chance to clean a stain properly, one useful trick I was taught during my school needlework lessons is to rub the piece of fabric housing the stain against another part of the same garment. It doesn't always work, but often does. Watch out to not be too vigorous with the rubbing though and so introduce loads of creases into the fabric. But you can do a bit and see whether it is working before carrying on. I've had great success with this method when the stain has come from splash-making wash basins in public toilets, and therefore just consists of water and a bit of soap.

Conversely, those public toilets can be very useful for sorting out spillages on clothing if you have to run them under a tap. Just be sure to avoid getting fabrics made from artificial fibres too close to the heat from wall-mounted hand dryers. Oh, and remember that other people also use these facilities and so you probably don't want to be stripping down to your underwear just to try and get a stain out.

Black Tie Events

Of course all these considerations take on extra meaning if you are attending an awards ceremony, for which you may be required to wear full evening dress. Here are some of the things I've learned after being lucky enough to be shortlisted for—and in a couple of cases win—awards.

4

If you are wearing feathers as part of your outfit, be prepared for some shedding both on the train travelling there, and during the event. (Leaving a trail of feathers as you exit a room may or may not be the lasting impression you want to give—that's a matter for personal judgement.) I have also noted a tendency for friends and colleagues who have had a bit too much to drink to want to try on parts of my outfits involving feathers. I'm not sure if this is universal, or just the people I happen to know! But it's worth noting that you may attract some unexpected attention.

For attending long work events I'd generally say that comfort is key. But when it comes to a black tie event I tend to focus mainly on what the outfit looks like. Turning heads is surely the primary aim, and you don't need to be comfortable, right? Actually, hang on a minute! You do need to be able to fit that five-course meal into your waistband. Plus, of course, it helps to be able to breathe! I was horrified to see the marks around the neck of a friend who had accompanied me to an award ceremony when he unbuttoned the collar of his shirt on the journey home. He said it had appeared to fit OK when he first tried on the borrowed outfit. But as the evening progressed, he felt increasingly strangled by the collar. So while it is a very good idea to hire evening outfits in terms of it being a sustainable and economical option, please don't make the same mistake of going for a size too small. Take your time with trying on, or if you are ordering without being able to try it first allow a bit of leeway for comfort and movement.

> You do need to be able to fit that five-course meal into your waistband

Personal Grooming for the Time-Poor

Whatever the event, and whatever you choose to wear, as mentioned earlier, paying attention to your personal grooming so you feel as confident as possible is extremely helpful. Of course, there are some people for whom a deliberately unkempt, scruffy image is one of the keys to their success. But for the rest of us, trying to look as tidy and professional as we can does help.

One thing I've learned the hard way is to apply the old adage of "little and often." Near the start of my career, I would often leave filing my nails and the host of other weekly grooming tasks that I carry out until a couple of days before a big event. By so doing, I was burdening myself with hours of

FIGURE 4.3
Tackling grooming tasks with a 'little and often' approach can help avoid catastrophes when getting ready for major events. (Shutterstock ID: 575609041.)

personal preparation that needed completing almost in one session. I kidded myself that I was behind on such things because I was so busy working. But the fact was that it actually took longer doing a massive 'catch-up', which then also felt like a real chore and became something I dreaded (Figure 4.3).

A friend kindly suggested that to avoid this I could set about instigating a pattern of doing a bit every day, and to look on it more as 'me-time'—a nice bit of pampering before or after a hard day at work. I took their advice and have never looked back. Sure, there are times when I get a bit behind schedule. But overall I am almost ready-to-go whatever invitations come my way, and that is definitely a less stressful situation to be in. In order to achieve this, not only do you have to summon the will to make that change, you also need to experiment with products and techniques that work for you. But to get you started I've got a few ideas and recommendations.

First, be realistic about the time you have available to spend. If you have children or caring responsibilities, for instance, or even a very demanding pet, you may not have as much time as you would like to spend on yourself. That's where multi-functional products come into their own, as does a bit of careful multi-tasking. The former might include products like a lip balm that also adds a hint of colour, a beard cream with built-in moisturiser or scent, or a combined shampoo and conditioner for hair. In terms of multi-tasking, this might involve cleaning your teeth while your fragrance or body lotion is drying, or filing nails while you listen to the TV news.

> Be realistic about the time you have available to spend

4

You can also look out for products that absorb or dry faster than others. I've certainly found quick-dry nail polishes an extremely useful addition to my bathroom cabinet, for instance. Having a polish that is touch-dry in a few minutes, as opposed to barely being able to handle anything for about half an hour makes such a difference, as do 'one coat' polishes. If I have longer it is nicer to use more traditional products, which do sometimes give better results. But knowing you have products that are quick to use in situations where time is tight makes all the difference.

If the only chance you get to think this through is when you've taken a holiday from work, then allocating the first rainy day to shopping for, or trying out, products is time well spent. If you find something you like, buy a couple more and keep them in the bottom of your wardrobe so you are not in a last minute panic when your perfume/moisturiser/sunblock runs out. I regularly check my 'stash' to make sure I top it up once I am down to the last reserve of a product that I rely on regularly. All this may seem like loads of planning, but it really can save a lot of stress, and thereby help you communicate to the best of your ability, when called upon.

5

Using Remote Technologies

Technical Support

5

"Oh, bleep!*" I exclaimed when I thought I'd just switched off something that had finally decided to work in my video conferencing software. This was while I was struggling to connect with a colleague, and I'd assumed my microphone was still turned off. But I heard a chuckle at the other end. "Well I can hear you now!", laughed my chum, revealing that my swear word was the first thing that had come through their speakers. (*Not the word I used!)

This occurred during the second wave of the COVID-19 pandemic in the UK, when I was trying out some unfamiliar software ahead of using it for some teaching. My kind friend and colleague had offered to give me a practice run with the software, and over the next half hour we investigated the

 DOI: 10.1201/9781003206828-5

program's features. This session ended up helping us both. So, I'd certainly advise you take the time to do a practice run with unfamiliar software (minus the swearing of course) before you need to do anything time-critical using remote technologies.

It seems to be Murphy's Law that every organisation or individual uses a different video conferencing system. Therefore, while

> Take the time to do a practice run with unfamiliar software

there is some commonality when it comes to looking professional during remote communication, which we will be covering in this chapter, the various software packages do still have their individual nuances. Consequently in order for remote meetings, talks, or any other professional events to go smoothly, you do need to be familiar in advance with these differences. On the day of your talk, meeting, or event, it is also worth booting up tablets or laptops well ahead of time just in case they need any software updates.

A Changing Workplace

Prior to the pandemic, it is fair to say that using online technologies for attending meetings and conferences, for teaching, and for delivering presentations was only slowly gaining traction. But by the summer of 2020, most of us had experienced a sea change in how we communicated. Meetings looking like the scenario shown in Figure 5.1 had become commonplace.

As I write this book, for some people the pandemic lockdowns and travel restrictions have left a legacy of remote or hybrid working. While most of us would not actively choose remote technologies over face-to-face communication, there are some advantages aside from keeping participants safe from any viruses. These include being able to give presentations from anywhere in the world without having to travel miles to do so.

5

Seeing Eye-to-Eye

One quite amusing though rather unpleasant aspect of using online technologies is regularly seeing the view up the nose of the person speaking. This is generally caused by positioning a laptop or tablet such that the inbuilt webcam is below the user's face. To help lessen this effect, try putting your laptop on a stand, or even on top of a pile of books. If you can, line up the webcam such that you are gazing directly into it when you look straight ahead—as I am doing in Figure 5.2a.

FIGURE 5.1
Work meetings via video conferencing software became the norm during the COVID-19 pandemic, and look set to remain a feature of our working lives. (Shutterstock ID: 1694685136.)

Looking into a webcam when you speak can be easier said than done though, because it is really tempting to look at the audience, interviewer, or your colleagues on the screen—especially if you are trying to gauge their reaction. We easily forget that if we do this, anyone on the receiving end sees you looking downwards—as I am in Figure 5.2b when my attention is focussed on the screen. So while you will of course want to glance at colleagues or audiences, you need to constantly remind yourself to look into the lens of the webcam so that it feels similar to how you would communicate if you were all in the same room. Otherwise, the people you are speaking to can start to feel quite disconnected from you.

If you want your hands in shot, especially if you want to demonstrate something, it is important to check that you are holding them up in such a way that

FIGURE 5.2
In these stills taken using video conferencing software, I am looking directly into my webcam in screenshot (a), while in (b) I am looking at my monitor. Looking into the webcam so that the people viewing you feel like you're in the same room as them will help you engage with them better than if you keep looking at their image on your screen. (© Sharon Ann Holgate 2021.)

5

they are being seen by the viewer. You could try recording yourself ahead of any meetings or talks to make sure you know where best to hold your hands.

Ready for My Close-Up

Another important thing to remember when communicating online is to dress your bottom half! I know the footage of the newsreader on Good Morning America back in April 2020 was hilarious when the camera angle moved to reveal his bare leg while he was clad in a jacket and shirt on his top half. But you really don't want to follow suit. (Or not suit as it happened in this case!) You never know if you might suddenly need to get up to retrieve something from a shelf, or answer the doorbell (Figure 5.3). Really messy backgrounds can also be very distracting for viewers, so if your room looks like a rubbish tip, just move stuff temporarily out of shot.

Dress your bottom half!

Sound can also be distracting, so if you have to work in a shared space, or in noisy surroundings, then it is advisable to use headphones. In an especially noisy working environment, it may even be worth considering wearing noise-cancelling on-ear headphones to help block out ambient noise. They are equally useful when you do not want the other participants' voices overheard.

FIGURE 5.3
However late you may be running, don't be tempted to take video calls dressed like this, just in case you need to stand up and move around during the call.
(Shutterstock ID: 1153859644.)

5

If you are working from home, while it can make for some amusing clips on YouTube, do try to organise things such that noises from your home life do not intrude on your talk. Maybe make a 'Do Not Disturb' sign for outside your door, or shut the cat or dog in another room with a toy to keep them occupied. Although maybe not a toy with a bell or one that squeaks! Also remember to turn off your phone or any other devices that could ring or make an 'alert' sound.

Keeping Switched On

While getting to grips with the technical aspects of remote technologies is of course vital, how you communicate via those technologies is equally important. Actor, facilitator, and communications trainer Gary Bates, who has worked for multi-national corporations including companies in the science and engineering sectors, feels that we need to work extra hard to build a rapport when communicating via video link.

He always takes time to chat to new clients outside formal meeting settings, and find out a bit about them as individuals. "When we used to be in a room with people it was easier because you would have those moments where you can chat in the lift, or in the corridor, or at the beginning of the meeting when you were pouring the tea out. You have to be more proactive about it when it's virtual, I think, and take that time. So I will often say to people: Where are you based? Or, where are you calling in from today? That kind of question can reveal a huge amount about people," he explains.

When giving presentations or delivering training via video, Gary says interaction is important. "It's so easy for people to switch off. So the more interactive you can make it the better. Ask questions, or ask people to put stuff in the chat. Have a poll. Try and get people to talk if it's an environment where you can get people to verbalise," he says, explaining that the training work he does is highly interactive. "On video especially people crave that because there's nothing worse than joining a Zoom session and being talked at for an hour and a half while someone goes through hundreds of slides."

> The more interactive you can make it the better

"Even if you talk for two minutes people can start to switch off. But if you then say to people: 'I'm going to put a question in the chat and I'd love to get your thoughts on that' it will make them think, oh, I should pay attention [as] I've got to do something," continues Gary, adding that he sometimes uses software to produce a word cloud with the participants' one word responses to a question. This, he says, is useful for highlighting the most common responses and is another way of keeping people's attention.

5

"In smaller groups, asking people to speak keeps them engaged. If people know you are going to ask them stuff and that they will be expected to speak they are not going to go off and check their emails and watch YouTube," concludes Gary.

Remote TV Interviews

The direction in which people are looking can often be a good indicator of how much attention they are paying. Most of us will, at some point, have been at lunch or in a meeting with someone whose gaze has wandered mid-conversation as they have become distracted by something. If this happens when they should have been focussing their attention on us it can be quite disconcerting. But it is equally disconcerting being intently stared at. Both extremes are best avoided not only in person, but also when we appear on television via a remote video link. This is because if we are looking straight into the camera, from the viewer's perspective it is as if we are talking directly to them.

To keep this looking as natural as possible, newspaper columnist and political commentator Andrew Fisher, who regularly appears on politics and current affairs TV programmes, recommends that while collecting your thoughts before replying to an interview question you briefly look away. You can then turn your gaze back into the camera once you start speaking, he says. "You don't want to be staring into the camera all the time. That's worse than looking away and then back into the camera again," he feels.

> You don't want to be staring into the camera all the time

5

All the Gear

Equally, if you are recording yourself speaking, it is a good idea to strike a balance between looking into the webcam or camera lens and occasionally glancing away. This can be facilitated by pulling curtains or blinds if sun is in your eyes, as you don't want to be squinting or in discomfort. In fact decent lighting conditions are important as you don't want to end up with shadow obscuring all or part of your face either. So if you decide that you want start your own YouTube channel, or want to look as professional as possible during online meetings or media interviews, you might opt to buy a selfie ring light. These are, fortunately, an inexpensive purchase, as are most tripods, and can make a big difference.There is definitely no need to have an enormous budget to record effective presentations or YouTube videos. By choosing to film at a time of day when the natural light is best, for example, you can get a good result whether you are using budget equipment or top-of-the-range technology (Figure 5.4).

You can get a good result whether you are using budget equipment or top-of-the-range technology

FIGURE 5.4
While an expensive kit like that shown in (a) is great if you have access to it, your smartphone and an inexpensive tripod (b) are all the equipment you need to get started with vlogging. (Shutterstock ID: (a) 1486167011 ID: (b) 1835655001.)

5

It is also well worth checking what features your existing smartphone or camera has. I, for instance, managed to film the YouTube content to accompany my textbook *Understanding Solid State Physics* on my compact digital camera. I was initially horrified to be unexpectedly faced with this task during one of the pandemic lockdowns when the professional lined up to shoot my videos could not enter the house. But I soon discovered that my camera could record in Full HD, and by mail ordering a selfie ring light and mini tripod, and carefully positioning my kit on a pile of physics books, I was able to produce a professional-looking result. So before parting with any cash, check carefully to see if you can make use of any equipment you already have.

5

6

Preparing for Media Interviews, Talks, and Poster Sessions

In at the Deep End

I could feel the start of flu coming on. I felt awful. So I decided to stop work early for the day. Just as I was shutting down my computer, I spotted a message from a radio producer asking me to appear on a show later that day with an audience of five and a half million. Thankfully, I made the choice to soldier on. It was a great experience, and the adrenaline kept me going. Years later, I got one of my favourite commissions to date as a result of someone hearing me on that show, and remembering me years later.

DOI: 10.1201/9781003206828-6

As this illustrates, there is sometimes little warning for media appearances, and hence very little preparation time. So in this chapter we will look at ways of getting yourself prepared quickly such that you can cope with a sudden broadcast request coming out of the blue.

Do Your Homework

In terms of getting ready for a radio interview, hopefully the producer or presenter will help you with the preparation, says radio producer of 30 years Dr Julian Mayers, who has made numerous programmes for the BBC and has a doctorate in astrophysics. "It's very much on them to say to the interviewee: 'this is the programme, here's how you can listen to a previous episode, here's the style of the interview and how long you are going to be on for'," he explains. If they don't send this information, then either request it or research it yourself, advises Julian. "You need to get a sense of what style and what length [the show is] and who the programme is aimed at. If it is a programme aimed at scientists, then that's one level. If it's aimed at laypeople, that's another level. So get a sense of who the audience is, and find out whether it is pre-recorded or live," says Julian, stressing that knowing this is critical even if you are asked to appear on a broadcast interview at the last minute and have almost no time to prepare.

If you do have a chance to, he recommends practising with a friend or colleague beforehand, answering their questions in an engaging and informative yet informal kind of way. "A good rule of thumb if you don't know your audience [background], is to tailor your answers as though they are an intelligent A-level [17–18 year old] student with an interest in science, or to imagine you've been asked what you do for a job at a dinner party," he says.

Dumbing down too much can be as bad as being too high-brow, feels Julian. "It's a fine line. You don't have to explain what an atom is if you are talking about physics. In the same way that if you are a politician you don't have to explain who the political parties are. You don't have to go to first principles," he says.

> Imagine you've been asked what you do for a job at a dinner party

Science topics can be complex though, acknowledges Julian. "Sometimes we're explaining quite difficult things and there is no harm in trying to find an everyday analogy—even if it isn't quite right. Analogies can work nicely as long as people understand they are analogies and that analogy can only take us so far."

The more scientists can humanise science and show that it's part of our everyday world the better, according to Julian. "Your job as a scientist is to make it understandable to the general public because in most cases the general public is funding what you do," Julian points out.

Cramming Revision

It goes without saying that if you are going to be broadcasting on radio, TV, or online about science you need to get your facts right. But you must also strike a careful balance between preparing adequately and over-preparing, advises Prof Peter Main, Emeritus Professor of Physics at King's College London in the UK, who has frequently been interviewed by media outlets both in the UK and internationally.

Peter has wide ranging experience of different types of radio and TV interviews, having discussed his own and other people's scientific research on air as an academic and while he was working as Director of Education and Science at the UK's Institute of Physics. The latter position involved participating in a diverse range of interviews discussing topics ranging from science policy, to nuclear power, to the representation of women in physics.

No matter what type of interview you are asked to participate in, Peter feels the primary thing that you need to do is identify your audience and then use appropriate language, avoiding using jargon. "If you're on TV the audience will almost always be the intelligent layperson as the people who don't watch the news probably won't be interested in science issues. But most people watching won't be able to understand jargon or know any background [to the science]," he says.

This is especially important to bear in mind when you're talking about your own research. "You are so familiar with your work that it's very hard to put yourself in the shoes of another person who doesn't know anything about it. It's even difficult to talk [about it] to other scientists sometimes," acknowledges Peter. (See Chapter 13 for advice on talking about your area of science with colleagues, and Chapter 2 for tips on how to identify your target audience.)

But whether you will be speaking about your own research or someone else's, and no matter what format the programme has, the focal point for your preparation for the science content needs to be the same.

"Decide what your main points are and be as concise as you can, because if you are on TV and radio you are going to have a very limited amount of time. Say a few sentences but don't drone on, and don't just answer 'yes' or 'no'. You must make sure that you get over what you want to say," advises Peter, adding that the interviewer will have a list of questions in front of them so can quiz you for more information if they need to.

It is equally important not to over-prepare though, feels Peter. "You need to think through what you are going to say, but don't prepare verbatim answers to questions, because then you lose all spontaneity. A good interview is a conversation," he states.

> Decide what your main points are and be as concise as you can

Mind the Gap

Unless you are being interviewed from home, you are likely to have to travel to one of the broadcaster's studios. If you have never sat inside a radio or TV studio before, this can be a daunting experience. Your voice and other noises sound completely different inside a closed radio studio, for example, and there can be a lot of equipment including cameras all around you. So imaging yourself being in the interview and coping well with it can really help.

I find that the journey to a venue can give valuable time for such visualisation. But don't leave all your preparation until then if you can avoid that. A noisy train carriage or aircraft cabin is not necessarily the ideal location for calmly working through your thoughts! However, it is better than nothing, and is incredibly valuable if you have had little notice time for the interview. Journeys can also be extremely useful for planning future science communication projects, so keep a notebook, or a phone that you can dictate into, close to hand for jotting down all your ideas (Figure 6.1).

> Imaging yourself being in the interview and coping well with it can really help

As long as you are not the person driving, if you are on the way to a media interview one thing to try is closing your eyes and imagining yourself in the scenario. Think through some different questions that might be asked, and imagine what you will answer. (In fact, visualisation can be a useful method of preparing for presentations too, as we will see later in this chapter.)

If the venue you will need to record at is near to where you are based, take a look at the outside of the building before the day of your interview. Alternatively, if you have to travel a long distance to the location, aim to arrive early so you can check it out ahead of time. This familiarises you with things like where the entrance is. That might sound like ridiculous advice, but on more than one occasion I have struggled to find the entrance of a

6

FIGURE 6.1
Time spent travelling won't be wasted if you can use it to visualise or plan your science communication.
(Graphic: Canva. Text © Sharon Ann Holgate.)

building and have almost been late arriving as a consequence. The last thing you want if you are feeling anxious about appearing on radio or TV is to get lost at the last minute!

There can also be complex checking-in procedures, so arriving in good time is essential. At the glossy London HQ of one publisher I worked for, I was staggered to find security so tight that having gone through several checks on the ground level, there were more security checks carried out when I reached the floor housing the meeting room. I'm not sure what they imagined could have happened in the journey up in a lift from one floor to another, but that was their procedure. Thank goodness I had not arrived at the last second!

Say What?

Just as important as turning up ahead of time for an interview is, of course, listening carefully to what you are being asked. This seems obvious and yet it is amazing how many people don't listen to the questions properly.

The last thing you want to do is appear evasive, or simply to fail to explain the science. Even if there is a particular message you need to get across, you must still ensure that you answer what the interviewer is asking. Otherwise the audience is unlikely to trust you because they will immediately spot any side-stepping of questions.

Unless you are speaking on the same topic very regularly, I would also suggest taking some sort of notes or prompts with you. For radio interviews I would highly recommend this. You have no idea

> Ensure you answer what the interviewer is asking you

what is going to happen on live broadcasts, particularly when there is a panel or group of guests. Debates can sometimes get heated if participants start to disagree with one another, or you could suddenly feel a burst of nerves. It is impossible to predict what will happen during a live broadcast. But having some notes in front of you minimises the chances of you panicking and completely losing track of what you were planning to say (Figure 6.2 and Box 6.1).

FIGURE 6.2
Taking notes or prompts with you can help you avoid going blank in broadcast interviews. (Shutterstock ID: 2170404441.)

6

Box 6.1   Preparing for Broadcast Interviews

- Find out the length, type, and style of interview
- Discover what the audience demographic is
- Think through what questions you might be asked and prepare answers
- Make some brief notes or prompt cards that you can take with you

---

Practice Makes Perfect

Software developer Rob Scovell, who has presented his work at conferences and meetings internationally, has some unusual advice for practising talks designed for small groups which he was first advised to do while studying at teacher training college. His then tutor suggested Rob go to the local laundry and give one of his presentations to the row of washing machines! Rather than startle other laundry customers, Rob now sometimes practices the technique on his kitchen appliances instead.

"Your flatmate or partner can easily get bored, and also you are just looking at them. So this technique is mainly to teach you to spread your eye contact around," explains Rob. "It's better than going through your presentation in your head. If you line up things and pretend that they are participants you then practice addressing each appliance. So when you come to do it for the [real life] group you are giving every participant equal eye contact. I know it sounds really weird, but it's actually really important if you want to get a good response from the group."

Actor, facilitator, and communications trainer Gary Bates, who has worked for multi-national corporations including companies in the science and engineering sectors, recommends practising your presentation by "saying the words out loud to yourself several times and recording it if you can. Use that [recording] as a way of editing and revising your talk and building confidence," he says.

"Then practise once or twice with a trusted friend or colleague if you can. The first time maybe sit with them, just talking at normal volume as if you were telling them this in the pub or in a café and trying to convey the story to them. Then the second time stand a little further away so it feels a bit more 'formal' and you need a little more volume and projection. But try to keep the same feeling as the first time," continues Gary.

6

Know Your Onions

Practice also involves revising your subject material so that you are completely on top of it. Rob stresses the importance of knowing your subject thoroughly, whether you are giving a conference presentation, or a much smaller talk to work colleagues or superiors. "It's so important to think over your material and really get the detail into your head. You've got to be able to bore for England before you go in there and talk about it. That way, you don't need to stand in front of a PowerPoint presentation and read bullet points," he says.

This ability can come in handy if you experience a tech failure during your talk—something Rob is familiar with.

> Think over your material and really get the detail into your head

"A few years ago I worked at a university in New Zealand and was sent to a conference in Australia with a laptop containing my presentation. The software was all running fine. So I'm there in a conference room with about 150 people listening, and the laptop just died 5 minutes into the presentation. But because it was in my own 'nerdzone' [Rob's shorthand for his area of expertise] I was able to just grab my marker pen and write stuff up on a whiteboard and talk about it," he recalls.

"I would advise to always have material in your head that you can talk about if your tech fails because conferences are notorious for technology failing," continues Rob. He suggests either taking a backup laptop, or converting your presentation into a PDF file and saving it onto a USB drive. That format is important, stresses Rob.

"Say you turn up with a Mac laptop and your presentation is in Keynote and it fails or can't be connected to the projector and the only other computer available is a PC. Or vice versa if you're using PowerPoint on a PC and there are only Macs there as a backup, you could be in trouble. But any computer will run a PDF, and even some projectors now have a USB slot that can run PDF presentations," he explains.

> Always have material in your head that you can talk about if your tech fails

It's All in Your Mind

Prof Richard Keegan, a sport psychologist and Professor of Sport, Exercise and Performance Psychology at the University of Canberra in Australia, insists that mental imagery should not just involve visual scenarios, such as going through the talk in your head and picturing the scene.

"I like to flesh a visualisation out beyond just vision to include smells, sounds and feelings. I really like to colour it in as richly as possible with every single sense. Layering everything up into your mind's eye appears to work better than just imagining the visual aspects," explains Richard, who in addition to his university role also works with international sportspeople and business clients advising on performance.

To help you get ready to cope should things ever go awry during your presentation, Richard recommends you imagine scenarios like the computer going wrong and how would you deal with this in front of your audience. "Having rehearsed plans in place makes you a bit more robust in those moments," he says. In general Richard finds around two-thirds of people have the ability to imagine scenarios like this accurately in their minds' eye. If you can do this, you can then use it to help you build a different future so that you don't live out the disaster scenario in real life. Those who can't easily do imagery often need to devote time to learning the skill before they can deploy it for a specific purpose, he adds (Figure 6.3).

But whether you are able to readily imagine like this or not, it is also really important to think through exactly what you are trying to achieve with your presentation, Richard explains. "Ask yourself: What is the one main message

FIGURE 6.3
To help you prepare, imagine the sights, smells, sounds, and feelings of giving your presentation, as well as thinking through any scenarios of things that could go wrong. (Shutterstock ID: 1909356121.)

6

or feeling I am trying to convey?" he advises. This clarity-of-purpose, he says, can help you steer presentations back on track if you can see the audience slipping away from you.

Having a very clear idea of what you're trying to do, coupled with plenty of practice—real and imagined—will help you get to a state which psychologists describe as 'flow', continues Richard. "With some of the best sports performances you ever see from people, if you ask them immediately afterwards how it felt they will say it felt effortless and natural and easy. It happens when you have this almost 'short-circuiting' between your goal and actually carrying out the task; skipping out the evaluative or self-conscious steps in between. But to get to that point there will have been a lot of practice, rehearsal and planning," he explains. So by practising and imagining your talk, the aim is to reach the position where you are closing the gap between your goal and your performance, and can intuitively navigate the feedback you're generating to stay on track: for example, via your audience's body language and engagement.

However, it is important to "remember that audience members will all have different expectations, and you may need to alter your delivery depending on what response you are getting," adds Richard, which is a skill he feels we can only really learn through experience.

There will, of course, be times when despite your best efforts with your preparation a talk starts to fall flat. In this case, you may have to opt to tough through it by carrying out what Christian Swann, one of Richard's former PhD students, has named a 'clutch' performance.

"When I'm talking to different athletes, I've discovered that they sometimes have to essentially grit their teeth and get through something," explains Richard. "So rather than your brain easing into this goal-feedback mode, it's this idea of 'fire in the belly and ice in the head': where someone is able to think in a very controlled and clear way. It's almost microscopic project management even if it's just for the next 10 seconds."

But pushing through can still lead to an excellent performance, continues Richard. "Some people will end up delivering a good presentation that way as well. They'll be looking at the audience and thinking about the message they want to convey, and they will be tailoring on the fly the next statement and the next interpretation, and making sure they're seeing nods of agreement."

(More tips on holding an audience's attention can be found in Chapter 7, while Chapter 13 gives some advice for dealing with a crisis of confidence during a presentation.)

Giving Talks

It's All in the Preparation

"I have such bad jet lag I barely know what's going on," the speaker I'd invited from overseas quietly confided in me while I helped them set up for their talk. Despite them being a leading academic in their field, what they said got me worried. But an hour later it was clear that I needn't have been concerned. They gave a fantastic talk. Their careful preparation saw them through, and I would defy anyone in that audience to have realised that the speaker felt anything other than 100%.

In Chapter 6, we heard advice on various ways to prepare for giving talks, and there is no doubt that being properly prepared is a key to giving a successful presentation. Unless you are incredibly lucky, being underprepared and slinging your talk together at the last minute will show. But as we will hear in

7

this chapter there is, of course, much more to delivering a top notch talk than preparing yourself and the technology you will use.

Addressing an Audience

When you are public speaking, one of the most important things to get right is the pace of your delivery. "Most people speak too quickly, so practising is useful to become aware of that. But you also don't want to become too slow and monotonous," says Gary Bates, an actor, facilitator, and communications trainer who has worked for multi-national corporations including companies in the science and engineering sectors. "You are aiming for a balance of clarity – people need to be able to hear and understand what you are saying, and take it in, especially with scientific information – but also naturalness. We all have our own natural rhythm and way of talking and you don't want to lose that. You are still you! So it's a case of adjusting your pace and tone while still being your human, authentic self."

"None of us is monotone in our real lives," continues Gary. "If you use story, humour, make it human, be authentic, and ask questions then your tone will naturally vary," he says.

In terms of vocal delivery, "you've got to have light and shade in your voice. You can't just drone on," echoes veteran magician and entertainer Mel Harvey, who has performed on TV, written and toured numerous theatre shows, and worked as a close-up magician in cabaret and on cruise ships for over 50 years. Mel feels that recording yourself speaking during your preparation for talks can help you avoid falling into a monotone trap and losing the audience's interest.

> You've got to have light and shade in your voice

Remaining "fixed to the spot can look robotic and become monotonous," continues Gary. "I always try to move when I am talking – at least some of the time. This can also help you connect to an audience, talking to and making eye contact with people on both sides of an auditorium. Moving can help you relax too," he adds, while acknowledging that moving can sometimes be challenging, for example if you are using a lectern and a fixed microphone. "Ideally you want a mix of stillness and movement," says Gary, who advocates keeping still when you are making important points. "Stillness conveys confidence, knowledge, and status," he states.

7

Keeping Nerves under Control

Actor and communications trainer Gary's top tip for managing stage fright is "Breathing! Before you go on, stand still and regulate your breath. Breathe in for [a count of] 4, then out for 4. Do this 5–10 times. You could also do some physical movement that helps you relax. For example, stretching your arms up as you breathe in and then letting them drop down on the out breath. Or swinging your arms round and round and loosely bending your knees as your arms swing down," he says.

"I sometimes find it useful to practise what I'm going to say while doing another simple physical activity, for example throwing and catching a ball or beanbag. This helps me to relax and takes the focus away from what I'm saying," he continues.

Getting into a positive mindset is also useful. "Visualise or remember when you felt positive or confident about something, when you had a success. Remember that feeling. Tell yourself that you are an expert, you know your stuff, and you deserve to be here. Remember that an audience is just a bunch of people—like you. They all have the same amount of insecurity, doubts, and worries as you. And they want to have a good time. They want to enjoy this experience. They are on your side, they want you to succeed," says Gary.

> Tell yourself that you are an expert, you know your stuff, and you deserve to be here

"Try to talk to an audience as an 'individual' – it's hard when you are looking at a sea of faces – but make eye contact with people and try to look at individuals. Talk to them as if you were talking to a friend and trying to explain the topic to them. Also having something to hold can help – for example a clicker, or even cards with your notes or bullets on," says Gary.

For veteran ventriloquist and actor Dawson Chance—whose past credits include his own TV series, three Royal Variety Performances, working internationally on cruise ships, and over 47 years of pantomime appearances in theatres throughout the UK—the best thing to do to help combat nerves is "to come across as confident. An audience can sense if you're not". Walking out with your head down or shuffling on is not a good start, says Dawson.

Even if you are feeling nervous, "just walk on confidently and engage the audience with a cheery smile. I always grab the audience in by saying hello and getting them to reply back. It's about starting to get some rapport straight away and getting them to engage with you," he explains.

If you are a naturally shy person, Dawson says you can learn to put on a persona in order to successfully perform or speak in public. "In show business it's called an 'act' because you have to put an act on. I've always been very shy, but when I go on stage I do believe in myself. So you do have to believe in yourself and that you can do what you're trying to do," he continues.

7

"When I was working on cruise ships we also had to host passengers on tables, and over a two week cruise we would have the

> Believe in yourself and that you can do what you're trying to do

same passengers with us every lunch and evening meal," explains Dawson, recalling that after one of his shows he got an insightful reaction from a husband and wife the next night at dinner. "They said that they really enjoyed my show last night, but that I was a completely different person! They asked: 'Why aren't you the same as when we sit down and have a meal?' I had to explain that this was because I was putting on an act. That's what I have to do to get my confidence up."

No matter what sort of venue or circumstances you are publicly speaking in, Dawson feels that if you don't have confidence in what you are saying the message will be lost. But how can any of us best go about faking confidence if we get a case of the jitters?

"One thing I would say that really helps with confidence is preparation," says Dawson, explaining that for past acting jobs he would use a lot of visualisation, going over and over in his mind every word of a script hours before a show. "I would lie down and picture everything in my mind: the moves that I would make, and what I would be saying. This was important to me to build my confidence up." Another good way of preparing is to practise what you intend doing in front of a mirror, he says.

Reeling Them In

"If you find some original idea to start the presentation, this will—even if it might appear a little bit bizarre—attract the audience's attention," says Mario Merola, Head of the Engineering Design Department at the ITER Organization in France, the world's largest fusion project. "I'll give you an example. Previously I was responsible for the ITER plasma facing components—so the components that directly face the thermonuclear plasma. These components are the first line of defence against the radiation heat that is coming out of the plasma, and so they are protecting the rest of the machine. I was asked to give a speech on this at a conference, about the plasma facing components of ITER, and on the first slide I put a medieval picture of an army that were fighting against some balls of fire that were being thrown from a catapult by some enemies. It was a scene from a battle. So I started with this picture and I could see that the audience was muted and thinking 'what is he doing?' Then I explained that this army is like a plasma facing component, and all the fire being thrown from the enemy is

7

like the heat coming from the plasma. The only main difference is that we love our enemy and we try to do all we can in order to preserve the plasma," he recounts.

"Maybe you are presentation number ten after a long day and so the audience are half asleep. So you need something to shake them awake," continues Mario, who regularly gives talks internationally to both specialists and the general public. He advises that holding people's attention is also about being a convincing speaker with a passion for the science.

"Clearly to be convincing, you must be convinced yourself. If you really believe in what you're saying, then your message automatically comes out in a convincing way. And if you are passionate in what you are doing, your passion and enthusiasm shows in your body language. For everyone other than professional actors, you can only show what you really think and believe. I am not an actor so I cannot simulate a feeling that I don't have!"

Gary feels the most important thing to do in order to hold the attention of an audience is to "tell a story rather than just state

> To be convincing, you must be convinced yourself

facts or information. Link it to your own experience if possible (or to an experience of someone you know, or even just someone you have heard about). By doing this you make it personal, make it human to you, and therefore to the listener or viewer."

"The best speakers – be they politicians, scientists, or business people – always create a story around the topic," he adds. "If you are able, a bit of humour is always good too. It not only adds variety and 'de-mystifies' complex topics, it also enables an audience to relax. Audiences are often nervous, unsure, and apprehensive. [So] it's reassuring to them to know that their speaker is human."

Depending on the context, getting interaction is always a good way to retain attention, continues Gary. "If people have drifted off, asking them a question or asking them to participate will quickly regain attention," he says (Figure 7.1).

But prevention is, of course, better than cure. So to avoid reaching the point where people do drift off, Gary has several suggestions. "If it's a lay audience, avoid technical jargon and language. Also avoid too many slides, and especially slides with lots of information and data," he says. "Visual slides are much better. Have an image, ideally an image which may intrigue the audience so they think 'what's that?', or 'what's that got to do with this?'. Then through your words or story you reveal

> Tell a story rather than just state facts or information

7

FIGURE 7.1
To help keep your audience as engaged as the one shown here, ask them questions and tell a story around the science you want to discuss. (Shutterstock ID: 1331491790.)

what the link is." Gary also cautions against "just talking 'at' an audience for too long", stressing again the importance of including some elements of audience participation.

Veteran theatre and TV magician and entertainer Mel Harvey, recommends scanning your gaze around the audience to keep everyone's attention. "Some people make the mistake of working to the first two rows. You can't do that and expect the whole audience to have an interest in what you're doing. You have to almost be on a one-to-one basis, and the way you do this is by looking in their direction individually," he advises, adding that even if the lighting means you can't see people clearly, looking towards them will still give the impression of talking to each person.

"You have to make a connection. The people who fail [at performing or speaking] are those who don't make a connection. If you're not interested in talking to them they're not interested in listening to you," Mel continues.

7

Despite your best efforts, audience members can still be distracted however. "These days you have to compete with people being on their phones all the time. I saw

If you're not interested in talking to them they're not interested in listening to you

it last night when I was working in a restaurant. People can't have a meal without a phone in their hand," says Mel, who often uses humour to defuse this difficult situation. For instance, he explains, he'll say something like "My talking is not interfering with your Zoom call is it?" to highlight the problem and make the other members of the audience laugh.

For more serious instances of phone use, for example when he doesn't want people filming his shows for copyright reasons, Mel makes sure that the venue requests people switch off their phones before the performance. He suggests making a similar announcement if you do not want your talk filmed, but stresses that unless you are happy to self-police phone use that you do need a staff member from the venue or conference team to police that policy.

Whether they are on their phones or look disengaged, if you feel you are fighting a losing battle to keep the audience interested try not to jump to conclusions, warns Dawson. "Someone in the front row could be sitting there with their arms folded. They might be enjoying it inwardly, but it's not showing on their face. So just keep calm and carry on."

They might be enjoying it inwardly, but it's not showing on their face

To keep the audience's attention, "you mustn't have any gaps in what you are doing. Everything has to flow," advises Mel. In some cases, people may drift off a bit if they cannot understand what you are saying, not because they are not interested, he says. "It's always worth being slow and steady with your delivery because not everyone in an audience listens at the same rate. Also, knowing your audience is an important thing because if you're too fast and English is not the first language for your audience, for instance, you may not be understood," continues Mel. (See Chapter 2 for more advice on delivering talks to overseas audiences.)

Dealing with Hecklers

Although we'd probably all like to think that scientists wouldn't interrupt a talk, I have certainly seen this happen multiple times. In one instance, the interruptions became extremely disruptive for the speaker, and I felt very uncomfortable sitting in the audience. This was because the questioning was, in my opinion, unnecessarily aggressive and the intention appeared to be to stop the speaker from delivering their talk by constantly rubbishing their arguments. So if we do find ourselves faced with a similar situation, are there any tips we can take from entertainment professionals for dealing with hecklers?

"I don't get too much of it, but if I do get hecklers I'm usually straight back in their face," states Mel, admitting that this stems from his time gigging in working men's clubs in the Midlands and north of England years ago, and is not something he would advocate when giving talks in a business or science setting. "But it is worth remembering that hecklers always forget one thing: that you are the one with the microphone!" If someone does interrupt and you want to use humour, Mel suggests "you could say something like: 'I would like to chat but I've got to do this!'".

> Hecklers always forget one thing: that you are the one with the microphone!

Veteran ventriloquist and actor Dawson Chance also recommends using gentle humour if possible, and admits that as part of his preparation before going on stage he has sometimes visualised dealing with hecklers. "You can get people trying to 'get one over on you'. They think they will, anyway! I've been fortunate not to have had much heckling, but when I have I've dealt with it through the [ventriloquist] dummy making a joke out of the situation. Humour is good to use if you can because you don't want to alienate the rest of the audience. You want to keep them on your side. So you don't want to belittle the heckler to the extent that you come across as very rude. You want the audience's sympathy" (Figure 7.2).

"As a facilitator rather than a presenter I always want to acknowledge the 'heckler' or the question. You don't want to put them down or make them feel unimportant or ridiculous because that could alienate your whole audience," says Gary.

If Gary has already planned to address a point raised, he thanks them for the question and lets them know the discussion is to come. If not, and he can readily answer he does, before checking if this had addressed their query. "If they say no, or ask another question then you may have to defer it and offer to discuss afterwards or in Q&A. If it's a moot point or if there is no obvious answer then either acknowledge that – thank them and say it's an interesting point and that we can come back to that afterwards. Or, if it's an environment where you want to engage the group in discussion and collective thinking, put it back to the audience. Ask them what they think," advises Gary.

Questions, Questions

Of course it helps to have agreed in advance with the session host how you want to handle questions. But unlike aspects such as travel arrangements

FIGURE 7.2
If you do get issues with someone heckling, using gentle humour can help to defuse the situation. (Shutterstock ID: 1803639517.)

which can be planned well ahead of time, things like agreeing the Q&A format, and explaining what recording and notes you are happy to allow, may only be possible to do a few minutes before starting your talk. But with all the other factors already prepared for, this should not faze you.

However if it does, or if you get a sudden attack of the jitters for any other reason, make sure you have your notes or prompt cards to hand in case you lose your thread once you begin an answer. However nervous you might feel though, don't ever be tempted to just read out the text on your slides or notes. Giving a presentation is a wonderful chance to expand on your ideas both during the talk itself and within the Q&A session. You don't want to reply to questions in such a way that the audience members may as well just read your slides again themselves and you are effectively redundant. (Chapter 13 contains more advice on dealing with feeling nervous before important communication events.)

One thing you may find helpful before delivering a presentation is to also ask yourself a few questions. When people have

Don't ever be tempted to just read out the text on your slides

7

sought my advice on combatting stage fright, it seems that for many their nerves are based on a worry about what the people listening will think of them. Personally, I don't feel that is a helpful train of thought when you're planning a talk. It is far better, in my view, to ask yourself which of your personality traits are best suited to the type of audience you are interacting with? You may be surprised what facets of your character, or what aspects of your hobbies you can draw on.

For instance, if you like R&B music and are due to address an audience of teenagers, you might base some examples on this, or speak how you would if you were discussing the music you love best. Conversely, you wouldn't speak like that if you needed to describe a complex piece of your research to your line manager in an industrial laboratory environment. So that is an aspect worth bearing in mind when anyone gives you that much repeated piece of advice of 'be yourself' when speaking publicly. Which version of yourself? Will it be the version that jokes around with mates in the pub? Or the version that details their PhD research in a job interview?

This all brings us back to the idea of tailoring to your audience, which was discussed in Chapter 3. But in this case we are talking about tailoring your own style of delivery, and considering what aspects of yourself will best relate to the audience. If you give this some thought, I am sure you will find a suitable approach—and more qualities and experience than you might have initially realised—to help you through your presentation. To conclude, Box 7.1 gives a little checklist of top tips.

Box 7.1   Top Tips for Delivering Talks

- Speak with light and shade, not in a monotonous tone
- Familarise yourself with the technology you will be using
- Make eye contact with audience members
- Start with a cheery smile and saying 'hello'
- Deal with disruptive audience members using humour

7

8

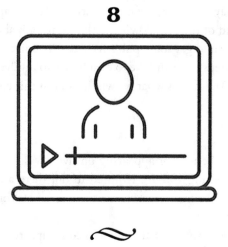

Radio, TV, and Online Broadcasting

Putting Your Head above the Parapet

I once spent the best part of an hour on the phone trying to persuade a scientist to appear on a radio programme that I was presenting. They seemed the perfect choice for the topic. I was really keen to interview them. They quizzed me endlessly on the aims of the programme, the audience demographic, and the topics that we would be covering. But it soon became clear that they didn't like the style I was aiming for. They turned me down on the grounds that I would "make them look stupid". Within the next hour I had another scientist lined up for interview, so by the time an email arrived from the original scientist saying they had changed their mind and really wanted to appear on the programme it was too late. They had been replaced by someone else.

Let me state for the record that it was, has never been, and never will be my job to make someone look stupid. What would I seek to gain by doing this? The whole idea of it makes no logical sense. My job actually involves doing quite the contrary. As we will hear later in this chapter, producers and

DOI: 10.1201/9781003206828-8

presenters tend to go out of their way to get the best from their interviewees so that the science being discussed is as clear as possible for the audience.

That is not to say that there are no potential pitfalls to appearing on radio or TV. But instead of turning down the chance to talk about your research to a wide audience and get some publicity for your university or institution, there are various steps you can take to enable you to say "yes" to that request with confidence. And that is what we will look at in this chapter.

In the Hot Seat

For many people, the idea of appearing on live TV fills them with dread. But if this is something you are required to do, there are various ways to help prevent it from being a negative experience. All of which require a little forward planning.

"Prepare. Think through what's going to come up," recommends newspaper columnist and political commentator Andrew Fisher, who from 2015 to 2019 was head of policy for past leader of the Labour Party in the UK Jeremy Corbyn. Andrew regularly appears as a guest on political and current affairs TV programmes, often with just a few hours' notice. TV interview durations are often equally tight for time. So Andrew advises against trying to cover too much information in one go, and being clear in advance what the main things are that you want to say, and how you can defend what you are saying if challenged by the interviewer or another guest.

"Never try to get more than three points across because there's never time to say more than that. Then for each of those three points think: 'Where will the interviewer find a hole in this, and what will be their follow-up question?,'" says Andrew, adding that you must ensure you have suitable answers ready (Figure 8.1).

> Never try to get more than three points across

If you don't know the answer to a question, be honest about that, continues Andrew. "That is even more necessary these days because when you're interviewed on TV, there will be a clip of it somewhere for eternity. You cannot afford to mess up, and it will follow you for the rest of your life!," he says, adding that he does not want to put anyone off from being interviewed, but is just trying to stress the importance of being properly prepared beforehand.

This preparation may need to include planning how best to cover the same ground over and over again for different broadcasts.

"If you have to do multiple interviews in one day it's very difficult to keep sounding fresh. So if there are slight differences in the audiences, you could

FIGURE 8.1
Don't try to get more than three points across in a TV interview. Also think in advance what the interviewer's follow-up questions to your points might be so you can prepare suitable answers. (Shutterstock ID: 2169462019.)

8

say a particular point for one audience and not for others. Or if you didn't manage to get a point across in the first interview, you could do it the next time," suggests Andrew.

He feels steering the conversation can also be a useful skill to develop depending on the type of science you are covering.

"During an interview if there are particular things you want to talk about, you can try making a link to another point so that it isn't just the interviewer guiding the interview. If the interviewer becomes quite combative you can see it as a jousting contest almost," explains Andrew, acknowledging that while political interviews have a tendency to get more heated than interviews about science, things can still become fractious depending on the research area you are speaking about.

If you're going to be talking about science that is contentious, and you've already made similar points on social media "you can look at the feeds to see how people have already attacked what you've said," recommends Andrew. "Then you'll get an idea of the sorts of criticism you may receive during the interview." He suggests you try countering arguments by explaining how you have come to those conclusions.

If you are on the receiving end of an unexpected or more generalised attack, Andrew suggests you point out that what you are talking about is something that you have studied for years. Then "give the science-based argument that leads to your point of view," he says.

> Give the science-based argument that leads to your point of view

Depending on the format of the programme, you could be one of several guests. So you may need to respond to their questions as well as those from an interviewer or from studio audience members. If you are part of a panel discussion, Andrew says it is important "to research the rest of the panel, so you're aware of their views and know what they might say. You can then plan your response."

Knowing the presenter's questions in advance also gives you a chance to plan your replies. So always ask the producer for a list of questions beforehand. Even having a rough guide as to what topics will need covering will help you in your preparation.

On the Receiving End

When it comes to conveying scientific information clearly and concisely on TV or radio, one of the most important things is to assume the audience has very little knowledge of the subject, feels Sony Radio Gold Award winning news journalist Owen Bennett-Jones. "Don't use any scientific terms, and no acronyms," advises Owen, who worked as a foreign correspondent for the BBC for many years, has taught journalism in universities in the US and the UAE, and whose current work includes presenting the podcast The Future Of… on the New Books Network.

If a soundbite is required for a given programme, he suggests pre-preparing this and making it approximately 20 seconds long for radio but only 10 seconds for TV. If you are invited onto TV this will, of course, involve you being seen and Owen recommends you "think about the setting and the clothes–[asking] what do they convey about you?"

As an interviewer, Owen finds "long, boring answers" problematic. "Scientists say they can't explain their ideas in a short time but in fact they can. It's just a question of using language that people understand and reducing replies to very broad-brush comments and principles," he continues. "That's all you are going to get across anyway. You are not trying to show off to avoid criticism from fellow scientists: you are trying to communicate to non-scientists." (Figure 8.2).

FIGURE 8.2
Taking time to prepare soundbites if needed, and ensuring you use everyday language that all listeners will be able to understand, will help make you an accomplished interviewee. (Shutterstock ID: 2092942681.)

8

When you are preparing for media interviews, it is more a question of thinking about how you are going to best explain the science than revising the science itself, according to Owen. "You should already be in command of the material, so don't worry about that. The difficulty is communicating your knowledge. Perhaps you could imagine a grandparent in your mind's eye and explain your subject to them. Or have a rule such as: don't use a word, phrase or acronym that a bright 16-year-old would not understand," he recommends.

> Don't use a word, phrase, or acronym that a bright 16-year-old would not understand

The best type of interviewee is "someone who is energetic, enthusiastic and above all else, clear," states Owen.

Don't Compromise

There could be times once a broadcast interview is underway where you feel a certain agenda is being adopted by your interviewer and you need to stand your ground.

8

"The media tends to reinforce simple messages, and can sometimes be a little bit contentious," feels Prof Richard Keegan, Professor of Sport, Exercise and Performance Psychology at the University of Canberra in Australia, who also works with international sportspeople and business clients advising on performance. He says he often hears journalists pushing for a simplistic but potentially controversial message when he has been interviewed for radio or TV. "So beforehand I'll try to get some sense of what an interview is about, and I'll ensure I am giving accurate information, of course, but also forcefully adding any caveats. So if I am talking about a topic such as anger management, for instance, I might say: 'you wouldn't apply this to every single person because we are all different.' I will really try to force those types of clarifications into the conversation."

On occasion, doing so has led to Richard not being used for comment, or content he has recorded not being broadcast. "But I would rather be giving accurate, carefully prepared information that won't cause harm—that is the absolute heart of my job. So I'd say that when you are discussing scientific information, always think about what your intent is, and what you are happy to be judged by," he says, adding that being registered or accredited (for example, as a psychologist) really makes this important.

> Think about what your intent is, and what you are happy to be judged by

Live in 3,2,1…

Once a TV interview is underway, Prof Peter Main, Emeritus Professor of Physics at King's College London in the UK and former Director of Education and Science at the UK's Institute of Physics, advises you to assume that your every move is being shown to the viewers.

"You are often surrounded by cameras and the producer will flick between them. As you don't know when you're on camera in a TV studio, don't start picking your nose!", says Peter, who has wide ranging experience of being interviewed on radio and TV by media outlets both in the UK and internationally. In particular, he feels it is worth being aware that shots taken over your shoulder from behind, as well as shots of the presenter reacting to what you are saying, might be used at different points in the interview if it is being pre-recorded and edited before transmission.

If you get a chance before the interview, particularly if it's being recorded, Peter suggests chatting with the interviewer beforehand so that you can be really clear what they want to get from you. One point worth bearing in mind, he says, is that "in general when it comes to science the interviewer is not trying to catch you out. So if you can develop a rapport with them that's good." If the interviewer is facing you, forget you are on TV and chat to them like you would to a friend, advises Peter, admitting that this ambience is trickier to achieve if you are remote in a satellite studio communicating via an earpiece.

But "they are trying to help you," continues Peter, adding that it is important once the interview is underway to be sensitive to your questioner's reactions. "You can usually tell if they've understood what you've said."

While he feels that as a scientist you rarely get the sort of aggressive questioning seen in some political interviews, Peter acknowledges that "the popular image of a scientist is that we know everything. But that's completely untrue." This assumption, he feels, can lead to a tendency for journalists to want a definitive answer, which is very difficult for a scientist to give. He has heard, for instance, scientists being pressed to give assurances that something is 100% safe.

"No scientist will say something is 100% safe. The key here is to reshape the question," says Peter, explaining that he would try to relate how safe something is to an everyday task that we all do without giving it a second thought even though that task carries a certain level of risk.

"The big difference between science interviews and other types of interviews is the jargon. Politics is complex, for instance, but there isn't a problem with jargon and you can simplify technical points using words that people can understand. But when you're explaining science there is both a jargon barrier and an understanding barrier. Even words like 'energy' and 'force' have baggage with them," he continues, adding that this makes the use of simple, everyday language when explaining complex science incredibly important. Otherwise, "explaining some of the more esoteric aspects of science is a bit like trying to sell a single malt whisky to a lager drinker. It has no meaning to them."

> When you're explaining science there is both a jargon barrier and an understanding barrier

Along with discussing risk factors, another tricky thing to do in media interviews is to debunk myths and conspiracy theories, feels Peter. This is a particular issue when it relates to public health, such as the conspiracy theory that mobile phone masts were spreading COVID. "It is difficult to dismiss the theory without sounding pompous and arrogant," says Peter. This attitude, he feels, can undermine public trust in science but equally so can sounding as though you're in doubt. "It's quite a difficult balance to be persuasive and not run the risk of sounding arrogant," he states, adding that it is imperative not to lose your temper if you're asked a question about something that from a scientific perspective is simply untrue. Instead, calmly explain why it is untrue.

If what you are going to be asked about is based on someone else's research make sure you read through and understand their scientific paper, advises Peter. "Even if it's your own research you need to think: what's the most important thing I need to get over about my research?" (Chapter 6 contains more advice on preparing for media interviews.)

Peter's last minute preparation just before going on air includes visualising in his mind what he will be asked, and finding somewhere to sit for a short while to clear his mind and breathe calmly.

"Go through in your head the things you've prepared, and expect to be nervous," he advises. "Don't think being nervous is bad. It's good because it pushes your adrenaline up and that adrenaline will help you think on your feet," he says, explaining that he was once called up to do a TV interview when he was breaking in a cold and found that the symptoms completely disappeared while he was on camera due to the adrenaline. "I started streaming again as soon as I left the television studio!" he chuckles. (Peter's top tips for being on TV are given in Box 8.1.)

> Don't think being nervous is bad

"If you like giving lectures, then you'll like being interviewed. Explaining things can be good fun. Once you are known as someone who can communicate well, the media will keep asking for you. So it can also lead to another strand to your career," concludes Peter.

Box 8.1   Top Tips for Being a Successful TV Interviewee

- Identify the audience
- Identify the points you want to make
- Make sure that you have in your mind answers to the most likely questions
- Look at the top of the camera, and don't be afraid to smile and look like you are enjoying yourself
- Try to keep as calm as possible
- Don't be afraid to pause before answering a question

Riding the Airwaves

A good radio interviewee "is someone who understands who the audience is and gives their answers or their part of the conversation according to that," says radio producer of 30 years Dr Julian Mayers, who has made numerous programmes for the BBC and has a doctorate in astrophysics. "So they are not dumbing down and they are not giving a university thesis. They are talking in an engaged, entertaining (if possible), and informal kind of way. They are not just reading out from a script. They are listening to the questions and they are answering in a pithy, but not necessarily a soundbite way. It should sound like it's an engaging, informed conversation. These are the interviewees you want to listen to."

It should sound like it's an engaging, informed conversation

"A bad interviewee is the opposite of that. They will not know what the level is. The worst are so focused on their own work that they will just talk about it as if they're giving a lecture to their undergraduates or to their peers. So they've made no account of who the audience is," he states. (Chapter 3 contains tips on how to identify your target audience.)

To help connect with the audience, Julian says a touch of humour is important. "If you can add in a bit of humour and humanity it shows that you are a human being as well," he says, stressing that this can help remove the "disconnect" between science and everyday life (Figure 8.3).

FIGURE 8.3
It helps to think of a radio interview as an informed yet informal conversation between you and the presenter. (Shutterstock ID: 160641761.)

"The audience can feel that science is 'out there' and it's not for them," he explains. But this gap can be breached if audiences can hear what a scientist is and what they do, and discover some of the process behind the science, feels Julian. To help achieve this, he suggests speaking about the challenges and questions that came up as you were carrying out your work, rather than just the results.

Admitting levels of uncertainty with your science is also important, he says. "We live in a world now where everything has to be black or white. But the world doesn't work like that," says Julian. "If we show people how science works, and that science is full of uncertainties then that helps the audience."

One of the things Julian strongly recommends is listening carefully to the questions being asked. He also warns to be patient with the interviewer. "It might be that the interviewer is not from a science background and asks what could be perceived as a stupid question, or asks the same question in a different way. Actually I'd say there's no such thing as a stupid question, because at the very least it makes you think about the basics [of the science]. And it may be the kind of question the public are thinking about as well," he says, adding that this type of questioning can therefore lead to the most insight for the listeners.

> There's no such thing as a stupid question

If a radio interview is being pre-recorded, it is particularly likely that you will be asked the same question again, says Julian. This is done "to make you sound good. So don't be offended if you're asked to repeat something." The producer may be looking for you to answer in a slightly different way, to explain something a bit more succinctly, want another quote to choose from if you said something they didn't quite understand, or have a fresh take if you stumbled over your reply, he explains.

Calming Your Nerves

The first thing I want to say here is that it is OK to be nervous! At its best, a radio or TV interview can be the perfect opportunity to talk about your work or a particular science message to a wide audience. At its worst, it can feel like a nightmare version of a school exam with, potentially, millions of people listening or watching and ready to pounce on you if you fail. So what advice does Julian have for coping with this potentially nail-biting situation?

If you are new to radio interviews, or tend to get very nervous, he recommends looking on it "as a conversation. Don't imagine millions of people listening. You are talking to one person and one person only—the presenter, or the researcher or the producer. So put out of your mind everyone else who might be listening," he advises.

If you haven't said something correctly, or you've stumbled over your answer, or think you could give a better response, then Julian suggests asking to reply to the question again. He also cautions against feeling compelled to keep talking.

8

"If you feel you've come to the end of your sentence then stop," says Julian, explaining that the presenter will soon fill the gap. "There's no onus on you to keep talking and fill up the airtime." When making programmes, sometimes there is no particular length of reply that Julian is looking for. "But 3, 4 or 5 minutes for a single answer: that's way too long," he warns.

Julian also recommends that if you're prone to feeling nervous take along a glass of water, as if we talk a lot our mouths tend to seize up, as they do when we're feeling under stress.

However, the best medicine against nerves, he feels, is to know what the level is and find out from the presenter or producer what type of questions will be asked. But don't be tempted to read out pre-prepared replies, he cautions. "Reading out just sounds awful!" says Julian, adding that it is also important to enjoy the experience. "It should be an enjoyable, entertaining, informal chat. No-one is going to die—I've not killed anyone yet with a radio interview!", he jokes. (For more advice on preparing for interviews, see Box 8.2 and Chapter 6.)

> It should be an enjoyable, entertaining, informal chat

Box 8.2   Top Tips for Being a Successful Radio Interviewee from Producer Julian Mayers

- Be succinct
- Be engaging
- Know the audience
- Relax
- Leave the audience wanting more and the programme wanting to have you back

---

Podcasts

Podcasting has come a long way since it first emerged in 2004. According to Daniel Ruby in his December 31 2022 article "39+ Podcast Statistics 2023 (Latest Trends & Infographics)" on the website of analytics and data platform

Demand Sage, there were 424.2 million listeners to podcasts globally through 2022, and at the start of 2023 over 5 million podcasts with a combined total of more than 70 million episodes were in existence. So podcasts have become an important part of the broadcast landscape.

As with any type of broadcasting, if you have been asked to appear as an interviewee on a podcast make sure you have heard at least one previous episode, and preferably more episodes. This will not only give you a feel for the duration and style of the interview but also what sort of level the questions are likely to be at. There is no point in preparing some excellent scientific puns for a serious podcast, and equally you might not want to be without them for a light hearted, entertainment-based podcast.

> Make sure you have heard at least one previous episode

Just as you would when preparing for interviews on radio and TV, it is also useful to ask in advance whether the podcast is to be recorded, or is going out live or 'as live'. Although with the latter there is generally not a huge amount of time to re-record anything, if you do make a bad slip-up there should be a small time window to do a re-take.

You may also need a technical rehearsal. Not everyone will record podcast interviews using the same software. So if you are invited to appear on a podcast, check beforehand whether you will need to install software on your computer. If so, make sure you do this with plenty of time to spare—if indeed you have the luxury of time. In my experience, IT issues always seem to happen when there is a rush to a deadline. (I'm sure that scientifically this can't be the case, but it always seems that way!)

When recording remotely I generally ask the producer if there will be 5 minutes available earlier in the day to test the connection. You may not get a 'yes', but there is no harm in asking. It can certainly help interviews to go more smoothly if they don't begin with a heap of stressful technical issues.

Podcast interviews tend to be a lot longer than those on radio or TV, so allow some extra time for revising and preparing what you are required to talk about. But although it takes longer to prepare for, the main plus side of a longer interview is that it generally enables you to go into more depth. You may get the chance to cover some of the more subtle nuances of scientific research that you might not get the time to discuss in other media broadcasts, for instance.

That said, rambling on or waffling around a topic is an absolute no-no. You still need to be concise in your explanations. It is just the case that you will hopefully get longer on air and so can say more things succinctly!

Newspapers, Magazines, and Books

Read Like You Mean It

"I don't think my students need to be told to read science stories," scoffed the academic at me when I was attempting to help advise them on science communication training. They had completely missed the point of what I was trying to say. As I then reiterated, I wasn't suggesting that the students weren't capable of doing background reading. I was recommending that they really analyse the structure, tone, and content of articles and how they differ from one publication to another. That sort of analysis is very different to casually reading something that interests you.

In this chapter, we will look at how to use preparation like that, as well as other tips, to help you write for the popular press. This type of writing is, of course, a very different prospect to writing an entire book. So in the second half of this chapter I will pass on some advice about things that have, and conversely have not, worked for me while I've grappled with the enormous task of getting a book over the finishing line to publication.

We will start by considering some fundamental ways in which writing a magazine or newspaper article is different from penning a research paper.

Not as We Know It

In fact the first step is simply to accept that, as discussed in Chapter 2, writing for newspapers or magazines has little in common with writing an article for a scientific journal. So if you are someone who thinks: "well of course I can write because writing is an integral part of being a scientist," I'd recommend changing your mindset.

To successfully write for a lay audience you will need to dip your toe into the waters occupied by professional science writers. Thinking that all types of writing are the

> Writing for newspapers or magazines has little in commom with writing an article for a scientific journal

same and that you have nothing to learn is what, in my view, holds a lot of academics back from being able to write well for the general public. A much more productive way of going about writing for newspapers and magazines for the first time is to look on it as a learning experience that will result in an expansion of your skillset. Just like any of your other Continuing Professional Development. Plus the good news is that you've already got off to a good start because I and the other expert I have spoken to for this chapter can help you with this upskilling!

In the House

There is no doubt in my mind that good writing starts with good reading. This is going to sound really obvious, but read whatever publication you intend writing for. (It's amazing how many people fail to do this properly!) Then read some competing magazines or newspapers. For each article, ask yourself some questions: Who was the likely target audience for that piece? What is the length of the article? Does it contain first hand quotes or just prose? Is there a political slant to the piece? How much historical reference is there? What is the overall style and tone?

This is a task that can't be rushed through in a few minutes. So perhaps set aside some coffee breaks for detailed reading (Figure 9.1). One thing that you will soon start to notice when you begin really analysing what you are

FIGURE 9.1
Set aside some time to really analyse what you are reading. Ask yourself questions such as who the article is aimed at, how long it is, what the tone and style are, and what types of content are included. (Shutterstock ID: 200247056.)

9

reading is that all newspapers and magazines have a "house style." This will encompass the punctuation and spelling conventions they use, as well as referencing styles, use of capital letters, and so on. You'll also find that publications have their own unique feel and tone.

Therefore while you will of course want to sound like 'you', whatever you write will need to fit with the house style of that publication. If you don't do much of that 'tailoring' work yourself you risk being very heavily edited, or in a worst case scenario your article may not be used. No one wants their hard work 'spiked', i.e. consigned to the waste bin. It is also not great to find your article extremely heavily edited. Both of these scenarios are much more likely to happen if what you hand in to your commissioning editor is very different to what they had in mind.

So how can you best avoid your articles requiring lots of editing, and cope with the editing process per se?

Being Edited

If you are new to a particular publication, my number one tip for reading the edited version of your article is to brace yourself! If it turns out to be much less traumatic than you are expecting, then that's a bonus. But it can be a brutal and humbling experience.

Before going any further, I should say that I value being edited. I know some writers who can't stand it at all. But it is nigh on impossible to proofread your own work. I do miss typos sometimes, as we all do. So it helps to have another set of eyes reading your work before publication. In fact, if you have the option of a friend or loved one reading your article before you submit it to the editor that would be worth doing as they will likely spot any small mistakes that you've made. Or indeed pull holes in your logic if you've not got a good flow to the piece! Both are extremely valuable types of feedback.

If that option is not available to you, try standing up or moving away from your normal desk, and read your article out loud. You'll be surprised how different it sounds once you can hear yourself saying it.

> It helps to have another set of eyes reading your work before publication

It is also worth making sure that you have a very clear brief from the editor, and that you query anything you are uncertain about. The editor may also be able to offer more generalised writing advice. To give you a head start, I asked former science magazine editor Dr Paul Parsons for some tips. In his previous career, Paul was managing editor of *BBC Sky at Night* magazine, following successful stints as editor of *Astronomy Now, Focus, and Modern Astronomer* and working as a features editor for *Frontiers* magazine. So what advice did he pass on to help writers with their commissions?

"Generally speaking, I was commissioning science writers, but there were occasions when we went straight to the expert for an article. The advice was the same - when commissioning anyone, I'd try to get them to be conversational in their writing. To almost (and this will vary depending on the publication, but was certainly true for *Focus and Frontiers*) imagine that they're explaining their work to someone in the pub. So, to keep things as simple as they can," says Paul, who is an advocate for drawing on metaphor and analogy when writing for the public.

"It's always better to speak figuratively and wave your hands a little if that gets the general idea across, than it is to get bogged down in technical detail which is going to go clean over the reader's head and leave them none the wiser. Keep it light, and don't be afraid to add a little flair to your writing now and again. This isn't a scientific paper, and most people will be reading it (in part at least) for entertainment," he stresses.

> Keep it light, and don't be afraid to add a little flair to your writing

A Stickler for Accuracy

Not delving into every technical detail can annoy some scientists: something I've regularly heard academic scientists grumbling about is that newspaper and magazine articles are 'inaccurate' or 'wrong'. Some of them then go on to start blaming the journalists who have written the pieces, complaining about their lack of understanding of the science. In my view, this stance is at best misguided, and at worst totally unfair. Let me explain why.

In some areas of science we are used to dealing with absolutes. If the solution to an equation is 2, it is 2. It is not roughly 2, or might be 3 because in someone's view it ought to be 3. However, a number could have a range of error associated with it. For example, it might be 2 plus or minus 0.1. This would need to be explained in an article about the work if this element of uncertainty is critical to the science, or its interpretation. But if it makes no difference whatsoever in any everyday sense, and only complicates the article for no reason, or would mean more important details being omitted due to lack of space, it should be left out.

9

Another issue can be with how scientists arrive at a numerical solution or a size in the first place. Getting to a result might have required incredible feats of accurate calculation or measurement that would simply not be encountered in everyday life. We don't tend to measure down to the nearest nanometre a space in our living room where we want a new bookcase to fit. That would be absurd. This level of detail is pointless and unnecessary in that scenario. Similarly, I don't need my room temperatures measured to tenths of a degree—even though I have some home thermometers that do so. I can even cope with a reasonably large margin of error on those measurements just for the purposes of adjusting my room heating. But if I were to be carrying out a temperature-critical experiment, or needed to store biological samples or vaccines below a certain temperature, suddenly the situation changes. I would need very accurate temperature measurements. They would be absolutely crucial to what I need to achieve.

So before describing numbers and sizes for newspaper and magazine articles, think carefully about what level of accuracy is required to accomplish the task of describing the science such that it can be understood by the widest range of readers possible. These articles are not designed to be read by experts in the field. Therefore, if you criticise the writing as being inaccurate, this sometimes fails to take into account how the article has been tailored to the readership. General readers will not want to know numbers to several decimal places, any more than you need to get out a micrometre screw gauge to check what width shoe laces you require—knowing roughly how big or small something is can at times be perfectly sufficient.

As we saw in Chapters 2 and 3, comparison with sizes of familiar things can help readers to quickly grasp the situation. However, as writers we do need to choose our analogies carefully. It might not be helpful to describe an atom as smaller than a football pitch, for example. That would be correct, but is an unhelpful comparison.

9

It is the same situation when it comes to including all the details of an experiment or result. When dealing with a tight space to fill that only a certain number of words can occupy, we are faced with a sometimes tricky choice as to what to include and what to leave out. At all times we need to carefully consider what level of detail is appropriate for the readership, while ensuring that any important nuances are not lost. Often it can be subtle points that are the crux of the story. So if you are a scientist being interviewed for a newspaper or magazine article it is important for you to point out any such nuances, and clearly impart to the journalist why this point is of such significance. No one wants to hear about how it took you several months to work out the best tolerance for a screw, for example. But they will want to hear when the rocket that screw helps to hold together is due for launch. OK, this is maybe a slightly extreme example. But you get my point. (See Chapters 6 and 8 for advice on being interviewed by journalists.)

"Excessive technical detail is a big no-no for me," continues Paul Parsons. "One thing that used to annoy me (from all writers, not just academics!) was trying to do

> Carefully consider what level of detail is appropriate for the readership

something clever, and failing dismally." This resulted in Paul having to spend time fixing the article when, he says, "all that was needed in the first place was a simple exposition of the facts. Remember the acronym KISS—keep it simple stupid!" He also dislikes "talking down to the reader" or using long, rambling sentences. "Try to keep sentences short and to the point. And avoid the passive voice - keep verbs active," he advises.

Then there is the issue of assumptions about the readership. Your potential readers may not have degrees in the area of science

> Keep sentences short and to the point

they are reading about. But they might have a degree in a related discipline, or in an arts subject. Or they may have no qualifications at all, but a keen interest in science. You simply don't know who is going to pick up that paper or magazine and read your article. But you can assume that if the reader has been bothered to start reading your piece, they are interested in that particular story or in science in general. This means they will likely have some basic knowledge of science and be familiar with things like cells and atoms.

"I think the best articles pre-empt what the reader wants to know, and then try to deliver that in the most straightforward terms possible," says Paul. "So try to put yourself in the shoes of the reader, and imagine what questions the subject you're explaining will raise in their minds. Then do your best to answer those questions."

While it won't always be obvious who your reader might be, most publications will have a good idea of their target readership, feels Paul. "Is it smart people who went to university but didn't study science? Is it blue collar workers? Is it school children? Find out who your audience is so that you can make your article speak to them. Your editor will be able to advise you more on this," he says, adding that anything you are feeling uncertain about should be discussed with your editor.

"If the brief's not clear enough, or you're unsure how to go about achieving what the editor has asked for, the best thing to do is seek clarification," states Paul. "At the end of the day, the editor wants the best article they can get for their publication. So it's in their absolute best interests to be accommodating and approachable to their contributors, and to help them deliver a great piece of writing."

Reviews

Depending on your job role, you may be asked to write a review of someone else's book for a journal, newspaper, or magazine. This is a completely different type of article to write than a piece explaining a science topic or talking about your own work.

One of the most important things to bear in mind when you're being asked for your opinion is that tastes can often be extremely personal. So it is worth considering how other readers may feel, alongside being honest about your own thoughts.

For instance, if the author has a sense of humour that you do not share this could make you dislike the writing style of their book, but that does not mean that the book is badly written. It might not suit you, but it could appeal to other readers. More generally, we all have our own preferences in terms of writing style, book layout, and graphics in just the same way that we have favourite radio and TV programmes. So you need to be really certain that your own personal tastes are not affecting your opinion and clouding your judgement about the science content. This may all sound terribly obvious, but it is in fact a very hard thing to do: it's not easy to step outside of your own thought processes and provide a really balanced opinion.

If you do not like a book for valid reasons, such as finding mistakes in the science or passages being incomprehensible or

> Be really certain that your own personal tastes are not affecting your opinion

badly ordered, then of course you must voice your concerns. I am not suggesting for one second that you do not give an honest critique. But the key is to be measured, accurate and constructive in your criticisms. It is a useful exercise to look at it from the author's perspective and consider how you would feel if you read this about your own work. You have no idea what sacrifices an author has had to make in their lives in order to complete their book, or what their circumstances are. Writing a review that is purely a hatchet job could lead to genuine distress, and may not reflect at all well on you.

You should certainly aim to quell any personal grudge that you may hold against the writer. Readers will soon sniff out petty jealousies and rivalries. You need to ask yourself if public attacks are what you want to be known for? If you have genuine criticisms, these ought to be backed up by factual evidence and proper reasoning, and delivered in a polite way. Think of it like feedback that the author could use to help pen a revised edition in the future.

Given the amount of editing that occurs during the production of most books, it should be rare that a book is truly bad. But if you find you have written a lot of valid criticism and are worried about how it is coming over, either read it aloud or get a friend or relative to read it and see what they think. Alternatively, leave the article overnight then come back and look at it afresh in the morning. If it sounds too brusque, you can tone down your word choices accordingly. In my view, there is no excuse for people being downright rude.

Former science magazine editor Dr Paul Parsons, who also worked as a science journalist and is the author of several popular science books feels "reviews and opinion pieces that require contributors to directly critique the work of their peers can be a minefield. I've known people who, upon discovering that the book they've been sent for review is a turkey, actually decline to write anything about it. They'd prefer to say nothing rather than say bad things and run the risk of whatever that could lead to. Caution is certainly merited."

"As an editor, I did once receive a threat of legal action from a researcher whose work was criticised in an opinion piece that the magazine had carried, and I know a book reviewer who has been furnished with a similar threat from a disgruntled author," continues Paul. "On the other hand, reviews of turkeys can be the most entertaining (and most useful) for readers. Editors know this, but at the same time they don't want a writ on their desk!"

"As a rule of thumb, truth is the ultimate defence. So be sure that you can back up everything you say with evidence—or at least an argument to support your comments that is solid in the eyes of the proverbial 'right-minded reader'. If in any doubt, ask your editor. If they can't help personally, some larger newspapers and magazines will sometimes have a legal department which will be more than happy to cast their eyes over your text for any potentially actionable remarks," explains Paul.

Another major thing to consider when writing a book review is what you would want to know about a book before purchas-

> If in any doubt, ask your editor

9

ing it or borrowing it from the library. Ask yourself: What are the main threads running through the book? Who is the book aimed at? Do you need any prior knowledge in order to understand the book? Is it an easy read, a proper page turner or really heavy going? The latter is not necessarily an issue depending on what the topic is. But you certainly need to analyse what category the book falls into. Is it something that you need to make notes on, or do further research to supplement it, in order to get a really good grasp of its contents? Or could you pop it in your suitcase and read it on your beach holiday? Depending on the publication you are writing for, and on the type of book you are reviewing, not all of these questions might be relevant. But you see the general idea.

No matter what type of book it is however, it is usually best to try to include a few direct quotes from it. Not least because this gives the readers of your review a good feel for the writing style and whether they may enjoy reading a book written in that way. Similarly, it is worth highlighting some of the topics that you found most interesting. But before doing so, check the house style of the publication that your article is destined for because there may be specific rules for things like quoting passages from books.

Books

If you really start to get the writing bug, you may consider penning a book yourself. This is the ninth book I have worked on (some written entirely by myself, others I contributed to) and my efforts have included popular science books and science textbooks aimed at various demographics. The best piece of advice I can give to anyone wanting to follow suit is: Don't! But if you really must, what are the main pitfalls to look out for?

The first is being unrealistic about the time writing a book will use up. I would advise thinking carefully about how long you estimate it will take you. Then double that time. Seriously. If you want to allow for any acquiring of images and for the editing phase, I'd say triple your original estimate. I cannot overstress how much we all underestimate what is involved in the task of completing a book. So please don't ignore the advice in this paragraph.

For most of us, becoming a book author will mean working into evenings and over weekends, and I want to give a real note of caution here. We only get one life, and it is so easy to miss out on precious moments with family and friends that you can never get back again. Loved ones do not live forever, and children grow up far more quickly than any of us would like to think. I really do urge everyone to think extremely carefully about how much of your personal life you are willing to miss out on in order to write a book.

A Place for Everything

If you decide that authoring a book is the right decision for you, then there are certainly ways to improve the experience. This is going to sound ridiculously mundane, but having got into a complete mess in my study when writing my first solo book, I now start new book projects by instigating a filing system. This covers both my PC,s hard drive and physical folders and notebooks.

Obviously, we all have preferred methods of organising our work. But whether you favour project books, a filing cabinet, or USB sticks, it is well worth taking a few hours at the start of your book project to think about how you want to store your research and your chapter drafts. Then there are all the things to consider that relate to your initial filing categories. These can include how you will tag topics that you may want to cross reference in various chapters, and how you will store your picture research and image permissions for associated photos, diagrams or graphics. I often find that I need to revise my filing categories as I go along. So every couple of months or so it is worth taking another look at how well your system is working, and making any adjustments as required.

However you decide to organise your material, you will need a bit of space to store it all in. A shoebox under the bed just

> Think about how you want to store your research and your chapter drafts

isn't going to cut it I'm afraid. So have a think about where this paperwork is all going to go as it increases in size during the project. That said, it is important to try and cull things like print-outs of drafts as you go along. I felt quite hypocritical as I typed the last sentence as I can be the worst offender when it comes to this! I tend to hang onto print-outs 'just in case' I decide to revert to an earlier way of saying things. But this leaves me with an enormous pile of sorting through and shredding at the end of a project, which is a situation that is really best avoided.

You may be someone who is a whizz at Excel and thrives on spreadsheets. If that's you, this could be the way to go - especially for checklists of things like getting in image permissions. Equally, you may be happier with a pen and paper. If so, don't underestimate simple schemes like using different coloured pens or highlighters to sort items, or to note when specific tasks or checks have been carried out.

It's really a case of trying to figure out what works best for you before you are two months into the writing and have a massive pile of documents in a complete mess (Figure 9.2). Or you become the possessor of an enormous computer folder that you dump everything in, and which soon becomes like a black hole for information. This is why I have said to not hesitate to change tactics as you progress with the project if some aspect of your filing system is not working as you had hoped. It is no one else's business how you organise your work, so make sure that it is optimised for you. You could need to look up specific things at very short notice, and you don't want to waste precious minutes searching for them.

9

FIGURE 9.2
Instigate a filing system at the earliest opportunity, and periodically check that it is still fit for purpose, to avoid getting in a complete mess with your drafts and other documentation. (Shutterstock ID: 1055690228.)

One final word of advice on filing is to remember to back everything up. I'm not happy until I have at least four digital copies locally plus one copy in cloud storage. Even then I occasionally lose things if I have a major IT issue. So this suggests to me that you can never have too many back-ups!

> Remember to back everything up

Making a Start

As with articles or frankly any type of writing, staring at a blank sheet of paper is of no use whatsoever. You will doubtless have had to supply the publisher with a synopsis or list of intended contents for your book and maybe even a sample chapter. So using this as your springboard, try to expand on what you already have typed up as quickly as possible. It does not matter if you need to return to what you write in these early stages and replace it. Just the action of getting down on paper as many of your thoughts as possible is incredibly valuable. It almost seems a sense of relief getting it out of your brain and into reality. But maybe that's just me! (Figure 9.3).

However it makes you feel, it is certainly a step forward and quite an important milestone. Not least because seeing more of your ideas laid down may lead you to slightly re-organise the content within the chapters, or even re-jig the structure of the book.

Don't be scared to make changes like that. The goal is to write the best book possible, and your publisher will want that too. There is generally likely to be quite a bit of room for manoeuvre if you start writing and realise that it does not make sense to follow your book proposal too rigidly. You can always consult your editor if you are uncertain about making any changes. They will be able to advise you. Certainly don't consider you have failed in some way if you need to make changes as you go along. Obviously, you have not done all the research by the point you submit the book proposal, so you need some degree of flexibility. There is simply no point in attempting to write the 'perfect' version of your book immediately. I don't know any writer who can do that.

> Don't consider you have failed in some way if you need to make changes as you go along

Keeping Up the Momentum

OK, so you are underway. You've had a bit of a tweak to your original book structure, and you are beginning to clock up some wordage. It's all going well, and then it hits you - a kind-of mid-term lull. You'd begun the project full of excitement and gusto, and you've put in so many hours you've

FIGURE 9.3
Six Stages of Writing

Text © Sharon Ann Holgate. Canva graphic.

lost count. But despite all of that effort there is still half of the book left to write. Plus you have hit a dead end on some of your research, and the number of picture permissions you need to get in seems to be growing by the hour. Oh, it's just getting too much! Will it ever be finished? Maybe I should never have even started the project!

And…breathe! You have started the project, and so now you will have to finish it. How hard can that be? It's quite tricky actually. Sadly, books don't tend to be published in part serial format like some were in the 19th century. But if you really think about that, it would be an even worse scenario. If you suddenly have a period of what we could loosely term 'writer's block', you may not have the next section ready at all!

9

No, there is nothing else for it but to grit your teeth and carry on. But I have found some ways of helping with this stage in the project. These all involve looking forward to a future in which your book is completed. One of the best methods I've found for cheering myself up and providing an extra bit of impetus for getting the project to the finish line is to start discussing ideas for the book's cover with my editor. Seeing draft covers from the art department is always very exciting, and it makes the book feel more 'real'. You can finally imagine holding the finished book in your hand. Suddenly you feel inspired again to drive on through the next few months to get the text completed!

Another pick-me-up can be to work on some of the picture research, if you have not already done so. Check with your editor whether your publisher has a licence with any of the photo libraries, or any agreements with other publishers to use their images. It can save a lot of time and hassle if you can make use of images that do not have complex licensing requirements. Plus in many cases you could be asked to pay from your own pocket if image rights need to be purchased. So unless you are completely obsessed by a specific image and there really is no other choice, I would always advocate going for free-to-use images.

> Grit your teeth and carry on

This could also be the moment to start thinking about creating any supplementary material that needs to accompany your book—such as a questions and answers manual for a textbook, for instance. The idea here is essentially to do something connected with your book that needs doing at some point, and will help provide a bit of a mental break from the main writing. That way, before you know it the chapters should gradually start appearing and the end will be in sight!

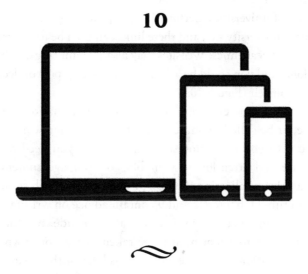

10

Social Media and Other Online Writing

Putting It Out There

"This website is eating my life!" moaned a chum who had volunteered to redesign and write content for a small voluntary organisation. I knew just how they were feeling as I was in the middle of a complete overhaul of my own website. The task felt never ending. In this chapter, I am relieved to say that I am not going to be giving advice on web design—this is not my area of expertise at all. But I am going to share experiences that both I and others have had when writing about science for online platforms, and pass on some tips for best practice.

Perhaps surprisingly, it is highly likely that writing online content will be part of many people's jobs within the scientific, medical, and engineering sectors. For many scientific institutions, their website is now the main 'shopfront' for their work. Although I (and doubtless some of my colleagues) have written part of the web content for some scientific institutions, not all of the webpages will have been written by science communication professionals.

DOI: 10.1201/9781003206828-10

These sites often include text provided by subject groups or individuals working or volunteering for the organisation.

In the case of universities, personal or group webpages are often linked to from the main university site, and these links can even be to the external personal websites of academics. Websites may also contain blogs, or blogs could be stand alone, and you could be asked to create a blog post or decide to take up blogging yourself (Figure 10.1).

These are just some examples from a range of scenarios in which you may choose, or be called upon, to write about your own work for publication online via websites or blog sites. In all cases, you need to apply just the same care and attention to your blog or webpage content that you would for a print publication. So don't expect to just dash something out in almost no time! Chapters 2 and 9 contain advice on writing clearly and succinctly, and on what content is essential to include. But there are also some additional factors that need to be taken into account when writing for online publication, which we will discuss later in this chapter. We will also look at various ways to create or improve an online presence, based on both my own experiences, and the research that I did for a couple of Science*Careers* articles (which are linked to in the Further Reading).

10

FIGURE 10.1
The start of a blog post that I was asked to write for the CRC Press website. This post trended on their website, but took me many hours to complete. So you need to be realistic about how long writing an effective post will take.

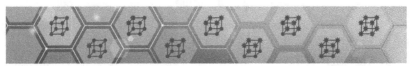

How solid state physics is helping us adapt to a COVID world

Posted on: April 25, 2022

Sharon Ann Holgate, science writer, broadcaster and author of Understanding Solid State Physics 2ⁿᵈ edition, ponders life without the technology based on solid state physics that we have all come to rely on during the pandemic

One day in the winter of 2021, during the third of the UK's national lockdowns, I was struggling to find the motivation to work amidst the challenges being brought by the pandemic. I found myself downing tools and sitting staring at the small tin on my desk. At first sight this looks unremarkable—a decorative tin full of paperclips. But I had bought that tin over a quarter of a century earlier for a very different purpose. Originally it housed the coins that I would save up and use to call home from University on a communal pay phone in my student accommodation block. All those years later, I found myself reflecting on just how much technology had changed the world since I went off to study in the late 1980s, and how much worse things would be for me, and so many other people, if the COVID-19 pandemic had hit us then.

The eighties was an era when almost no-one had a mobile phone, a home internet connection was unheard of, and video conferencing was something seen only in TV science fiction programmes. Even our family home computer was something I mostly used for playing a certain popular driving game, when I wasn't tinkering around learning some BASIC coding. The idea of remote working or teaching via that much-loved machine was laughable.

Almost a year after the epiphany triggered by my little tin, we are still in a world of blended learning and hybrid working that is

About the Author

Sharon Ann Holgate is an award-winning freelance science writer and broadcaster with a doctorate in physics from the University of Sussex in the UK, where she was a Visiting Fellow in Physics and Astronomy for nine years. Her credits include presenting on BBC Radio 4 and the BBC World Service and writing for Science and New Scientist.

Of course, you may also choose or be asked to communicate science via social media, and in this next section we will hear some tips for creating successful social media posts.

The First Post

Stepping onto social media for the first time is daunting for any of us. But it is important not to get too hung up about creating a post, feels Alexia Alexander Wight, an MSc physics student at King's College London who has previously worked as a physics teacher in a technical college, and is 'missneutrino' on Instagram and Twitter.

"One of my top tips is to stop being a perfectionist and just do it!" says Alexia, who has over 3,900 followers on Twitter and over 3,700 followers on Instagram. "Often my most successful posts are not the ones I've spent three hours planning. Some of my most popular posts on Instagram have been things I've done very quickly."

10

Alexia explains that while she aims to post three times a week and at the very least once a week, the type of content she puts out varies. For instance on Instagram she might share one of her Tweets, a

> Stop being a perfectionist and just do it

funny meme or comment about something she has just learned, or an illustration which has taken her hours to create.

"I would recommend having a range of content on your feed. On Instagram where I'm actively creating content I try to have a mixture of some photo-based posts, some Tweet screenshots, and then some longer, information dense posts. On average I take between half an hour to an hour on a post. But that means some of those posts take me fifteen minutes and some take three hours," she says. Alexia stresses that if you're always spending several hours on a post that might not pay off in terms of generating more engagement. Not least because "it's often the ones where I've posted something silly about something that's happened in class that has resonated with people," she says.

"Although on the one hand a lot of the time I put effort into something that doesn't pan out, that's not to say you should never make those chunkier, more information dense posts. One of my most popular posts on Instagram is one that I did spend a lot of time creating. It was talking about Cerenkov radiation and it got nearly two thousand likes, which is at least a factor of ten above the normal number of likes that I get," she continues, adding that in general her posts that get the most traction are those in which she gets really excited about physics. "My most successful post ever is a funny video of a graph. Making science approachable and human and talking about my lived experience is what is working."

Alexia admits that it took her about six months to develop an established style, which even then still evolved gradually over time. If you want to have a consistent style immediately, Alexia recommends using a graphic design software package like Canva. "Find a few post templates in Canva and stick with these to be really consistent," she advises. "I would highly recommend doing a series [of posts] as a way to try and build your style. Also, to check if your [planned] posts are likely to hit the mark it can be useful to talk to teenagers and younger people for advice."

What Am I Doing Here?

10

One aspect that can help you hone the types of posts you create is figuring out why you are actually on the platform, says Alexia, who has been hired for graphic design work after employers spotted her infographics and illustrations on social media. "You need to work out what you want out of social media. Are you trying to make social media your core [job] or are you building a network for future work? My focus now on Instagram is building up my portfolio. My ideal job is one foot in academia and one foot in science communication. So I'm doing posts partially to connect with the audience – so I do still think about whether people will like it, but I am also looking to practice science communication. It is really easy to get hung up on likes, but when I started to switch to viewing creating my Instagram content as building my portfolio it became a little less stressful. Building an audience is good, but if you're using it to help build your skills you don't need 10,000 followers" (Figure 10.2).

Whatever your reasons for being on social media, it is important to tailor your posts to your audience. Alexia mainly uses Twitter for commenting about science and her experiences of studying, as well as interacting with science communicators, academics, and other scientists more senior than herself. By contrast, her Instagram is more geared to teaching people about science and giving study tips, as most of her followers are undergraduate or postgraduate students. This, she explains, is partly due to the contrast in the age demographic of the two platforms.

"On Twitter there are a wider range of ages on the platform, whereas there are not many people over about the age of 35–40 on Instagram. The type of science communication I do on Twitter is more sharing my experiences as a student or as a person with a disability. I do less teaching of people on

FIGURE 10.2
It helps to consider what you are aiming to get out of being on social media. (Shutterstock ID: 1241714353.)

10

Twitter," says Alexia, joking that she would look "pretty silly" telling professors on Twitter where to read about particle physics, for instance.

She does, however, network on both platforms. She recalls how when she was a teacher this led to her securing a good speaker for giving a talk at her college. More recently, networking with CERN scientists via social media helped her build relationships more readily in person when she was on a summer programme at the research facility.

If you want to find your feet online before taking the plunge with posting, Alexia suggests that "you can try joining social media platforms anonymously. But this is easier on Twitter than on Instagram. Some of the most popular physicists on Twitter are completely anonymous. So they don't have their face or name on there, they just tweet. But that's obviously a problem if you're trying to use it for networking!"

Another tip from Alexia is to start a new account if you have been on a platform anonymously and now want to be public. This is so that your account gets flagged up as new.

Being Authentic and Consistent

If you decide to go public, "you need a persona on any social media platform," continues Alexia, who chooses to share some personal details online. "I'm definitely not the only autistic female physicist out there. But there aren't many who have visible online brands." This visibility, she says, makes her more accessible for younger scientists to reach out and engage with her. Trying to help them is important to Alexia who "found it really daunting going into physics and not always being able to see people with my experience. That is one of the reasons why I am open about being autistic and queer and disabled. It's not necessarily that I want to make that a focus all of the time."

No matter how much you want to encourage others, trying to be pleasant to everyone can only be taken so far though, feels Alexia. "If you are a young woman, be aware that especially on a platform like Instagram you will get people who are being weird and hitting on you. I would say do use 'block' or 'mute' on social media platforms if someone is continuously being weird. Don't be afraid to block people because you are trying to be nice and are in a science communication account. It is still alright to set your boundaries."

It is also OK to only have a handful of followers, because depending on the audience you are reaching this can still be influential, feels Alexia. She says that if by identifying with you it encourages even just one or two young people to build careers in science, being on social media will have been worthwhile. As she puts it, "you can only be seen if you put yourself out there."

> You can only be seen if you put yourself out there

FOMO (Fear of Missing Out)

Social media is a big part of many people's personal and professional lives. But if you are not using it to build a portfolio, what other advantages does being active on one or more of the social media platforms give scientists and science students?

First, it gets your voice out there so that people in your field of study know what you're working on and what you're about. It can therefore lead to all sorts of invitations, ranging from being asked to speak at meetings and conferences to prospective collaborations. It can also be useful if you're starting out in your career or looking for a new position, because recruiters can search social media platforms for job candidates. In addition, hiring faculty and managers could look at your public profile.

Some scientists have told me that they find social media ideal for pointing them towards research papers or funding sources that they otherwise might not have seen. Others use social media for public outreach, and of course it can be very effective for that. One project I was reading about recently in *Scientific American* was a Facebook Group set up by a herpetologist to help local residents differentiate venomous snakes from non-venomous and thereby prevent so many snakes being killed by his terrified neighbours.

However, like many things in life, social media is not entirely positive. Choosing to be on particular platforms—or possibly not to be on any at all—is a very personal decision which in my view takes careful thought. If you've ever found half an hour disappearing when you thought you'd only been flicking through someone's Instagram or Twitter feed for 5 minutes, you will realise how much time social media can suck up. This time sink issue is often compounded when creating your own posts. So you really need to think about how much time it is worth devoting to social media. (This was an issue I had to grapple with, as revealed later in this chapter in my Social Media Diary.)

> Choosing to be on particular platforms—or possibly not to be on any at all—is a very personal decision

10

Calm Yourself!

Certainly, getting into online spats can easily occupy a disproportionate amount of your day. These are also incredibly risky in terms of the potential for damaging your reputation or collaborative relationships. Another risk to collaborations can be if you paste up research results without your collaborators' blessing.

Equally, you can get into deep water legally if you post copyrighted content, results that have not yet been published, or information from conference sessions that the speaker does not want publicly disseminated. You can even jeopardise future patent applications if you talk about early stage research or inventions, or risk journals refusing to publish a paper based on that work.

Reducing the Likelihood of Problems

But although this all sounds very negative, there are strategies we can all use to mitigate these problems. Setting aside a measured block of time each day to do your social media can avoid the hours running away with you, for example. And if someone writes something you find unacceptable, take a bit of time out before replying—if indeed you decide to reply at all once you have given it more thought.

If you check with collaborators before blogging or posting about joint work, that should avoid potentially damaging that working relationship, or your professional reputation. In terms of future patents or journal publication, the key is to look very carefully at the regulations they stipulate before deciding whether or not to post something online.

Branding

Another aspect that is worth considering, especially if you intend trying to cultivate a large online audience, is branding. This might sound like something that only big companies need to worry about. But actually, even a single image of yourself posted online will create an impression about you—either good or bad, making it something that is worth carefully thinking about.

10

I've listed some ways to brainstorm your own branding in Box 10.1, and thereby help ensure you don't put off potential future employers or collaborators. Some of these questions are not easy to answer. But they should hopefully help you to understand what exactly it is that you're trying to achieve when you go online.

> Even a single image of yourself posted online will create an impression about you

This exercise should also help you to avoid compromising yourself in any way. I know that some people have two Facebook accounts, for instance. This allows them to post private pictures of nights out which they share with friends, while concurrently presenting a more carefully curated public face.

Box 10.1   Branding Brainstorm

- Describe yourself in three words
- What do you stand for?
- Who are you trying to reach?
- What impression do you want to give?
- Why is science important to you?

Taking your Time

If you've decided to go ahead and start showcasing your research or opinions to the world how can you best go about that? We saw some advice on this from Alexia earlier on in this chapter. But I want to add my own note of caution, which is particularly relevant for social media platforms where word-age is restricted: when you are posting anything other than a quick personal view take…your…time.

Double check your posts for factual accuracy, and check the copyright situation for anything you are intending putting online. Do not just steal content or images from other people's presentations, or from anywhere else on the Internet. You must have the rights to use the content you post. If in doubt, referring to the information in the Help Centres of the social media platforms will usually assist you. If you are still in doubt, play it safe and don't post it.

In terms of the science itself, allow enough time to ensure you've written in such a way that no caveats or context are

> Double check your posts for factual accuracy

lost. If some of what you are intending mentioning is less familiar to you, you could even consider running your content past an expert colleague in that research field before you post it.

10

It is also really important not to overhype the significance of either your work or that of others. Finally, take your time responding to any comments or queries: you don't want to have carefully fact checked your original post only to make a silly mistake further down the line.

Accompanying Images and Nice Backgrounds

If there is a stunning image that can accompany your post so much the better, because striking images will grab people's attention. Images can also be used to convey scientific information directly, in some cases via annotations. But beware of the different quirks of platforms. I soon discovered, for instance, that Instagram can automatically crop pictures. So if you are not very careful with what you click, you may not get the result you were expecting showing up in your posts.

If you wish to paste text-based posts, you may prefer to have a design as a background. Although there are other free software packages available online, I have opted for using Canva for some of my LinkedIn and Instagram posts. For example, the Instagram post shown here in Figure 10.3 is created using a free template within Canva. Sometimes I adapt my Instagram posts to suit re-posting on LinkedIn, and Canva is useful for making quick tailoring changes while still retaining the same overall feel and look.

10

MUCH AS I TRIED TO AVOID IT

my Easter looked a lot like this!

But surely I needed to replicate some recent scientific findings...

#HelpMe!

If you actually want anyone to read and engage with your social media posts, you are going to need to familiarise yourself with hashtags. These are essentially a way of grouping posts into topics so that people interested in that subject can find relevant posts more easily. If they choose to follow a particular hashtag, such as #brainresearch, they will see posts that contain this hashtag in their feed.

Trying to decide which hashtags to use on your posts is really daunting when you first start out on social media. But looking
Familiarise yourself with hashtags

up some hashtags via the platform's search facility, and investigating what hashtags people or organisations posting about similar topics to those you intend to cover should help push you in the right direction. Just be careful with what hashtags and content you choose to post, to avoid coming a cropper as in the incident I describe in my Social Media Diary (Box 10.2).

BOX 10.2   MY SOCIAL MEDIA DIARY

August 2, 2021: Well, the day is finally here! I've decided that I must create my first post today. How hard can that be? I'm really happy with the photo I took (Figure 10.4), so that feels like a great starting point. An hour and a half later…I'm still not quite ready to post, having written then deleted several versions. So I've decided to keep things really simple and not go into too many details about the flower. I have a small handful of friend's children who have all been on Instagram for years and have kindly agreed to follow me and give feedback on my posts while I learn how to create them. So I email them to say it is up, and see them duly requesting to follow me.

Mid-August 2021: I'm on Instagram completely anonymously at the moment, so I can get a feel for the platform. But although my account is marked 'private' and I've only posted a couple of times, to my astonishment I've received a request for someone to follow me.

10

FIGURE 10.4
The photo I took for my first social media post on my @everydaysciencethings Instagram feed.
(Photo and text © Sharon Ann Holgate.)

I hadn't planned to let anyone who I don't know follow until I am more confident posting, but I haven't said who I am so I've decided to accept the request out of curiosity. I'm beginning to start doing the occasional post, but these are a couple of weeks apart from one another due to pressure of work.

September 2021: I've been so busy with other things that it has been about a month since I last posted. Upsettingly, I've just had a look and seen that my one member of the public who had been following me has dropped away. It's really frustrating not to have even been able to hang onto one follower, but posts are taking me half a day to create and I just can't spare the time.

October 2021: I get chance to create a couple more posts, but don't think this is sustainable with the workload I have. I'm now wondering if maybe I need to admit defeat and give up with this?

April 2022: OK, so I've had a break of a few months and I have had a change of heart. The feedback from several friends' social media-savvy children on my early posts has been both positive and instructive. They are all encouraging me to continue on Instagram, so I've plucked up the courage to make the account public.

May 5, 2022: I post publicly for the first time (Figure 10.5) and get my first like. I stare at it on my tablet screen. Yes!!

July 2022: Finally I'm beginning to get into my stride, and I'm developing a 'style' with these posts. I'm still not posting as often as I would like, but I am slowly gaining followers.

August 2022: This month I'm going to aim to post at least once a fortnight, and to get used to factoring this into my working weeks. The rate at which followers are joining me is beginning to increase.

August 23, 2022: I've just created a post about some Roman walls that I recently saw on a visit to Chichester in the south of England. Each time I post, I've been checking that it appears in the hashtag feeds for the subjects I've been hashtagging at the end of my posts. It is good to see a couple of likes coming in almost immediately for this post, but I've just gone to look at the hashtag feed for #romanbritain and after the page refreshes my post has disappeared. Similarly, my post should be on #chichester and it just isn't there. What on earth is going on?

After searching online for advice I conclude that I have been 'shadow banned', which means my posts won't appear in hashtag feeds for a period of several days. My content about ancient Roman weaponry

FIGURE 10.5
Screen grab of my first public Instagram post @everydaysciencethings. (Photo and text © Sharon Ann Holgate.)

10

must have inadvertently fallen foul of guidelines on what you can post. It looks like it has been flagged as dangerous content. So I quickly edit out the paragraph that explains what the large tower-like structures in the Roman walls were actually for, and repost.

(I spend the next week panicking that my shadow ban isn't going to lift. But eventually it does and I start posting again and appearing in the hashtag feeds.)

January 18, 2023: I have reached a milestone with my first 100 followers, and I'm absolutely delighted! I am getting 'likes' on my posts and feeling glad that I decided to stick with creating this feed.

February 2023: I'm beginning to realise that it may not be very easy for me to develop a large following because of the range of subjects I write and broadcast about, and the variety of more personal subjects that I include in my posts. Because of this range, my appearances in subject hashtag feeds are quite scattergun compared with how they would look if I only posted pictures of wildlife and plants, for example. This will unfortunately reduce the chances of potential followers finding me. But I'm going to carry on as I am because I am starting to

@everydaysciencethings that does indicate a broad variety of topics are likely to be covered.

Spring 2023: I am now starting to gain followers at a rate of several each week and although, frustratingly, some of these seem to drop immediately away within about a day of joining, I'm really pleased with how the account is building. I'm also enjoying interacting with my followers and with the people I am following on the platform. I am very aware that I still have much to learn, not least in terms of how to best present my content, so there is still a long way to go. But I've found a way to fit this into my schedule, I am finding posting rewarding, and I am hoping that my followers continue to enjoy seeing the results of my efforts.

You can follow my evolving Instagram journey @everydaysciencethings

10

Spatial Awareness

Space may indeed be the final frontier. But you certainly don't want to push the concept of endless space on the Internet to its limits. While social media posts, by the very nature of the platforms they go on, have to be short, no stringent rules tend to apply to websites. So it is easy to assume that because you have a remit to create your research group's webpages, for instance, that you can paste up masses of information. In reality, institutions and companies are likely to have limits on how much content you can include, not least because they will be paying for web space.

Therefore, if you are asked for, say, a concise biography of yourself, or a short paragraph about your research, it needs to be exactly that. Short. When I have written content for the websites of scientific institutions, there have always been strict wordage limits. This means you may well have to edit down what you might originally have intended pasting up.

This is no bad thing. With the pace of life ever-increasing for so many of us, who has the time to read 30 webpages? You may want to create 30 webpages of content if there are different strands to your work and readers are most likely to access just a particular section that is of interest to them. But it is highly unlikely that any individual will look at all the content you post online.

It is also far better to post less content that has been carefully thought through than loads of poorly planned content, and to give yourself time to make your content as accessible as possible. The latter is something which is fortunately gaining increasing traction.

Access All Areas

"What's really nice online is there is a big trend on social media at the moment towards providing alternative text when you have text on pictures," says Dr Frankie Doddato, a Teaching Assistant in the physics department of Lancaster University in the UK and disability advocate. This text, she explains, gives a description of whatever words or pictures are in the post. People are also repeating the entire text in their main post with no distracting backgrounds. "That's really, really useful," enthuses Frankie.

I, for one, make sure I spend an extra few minutes to create alternative text (often known as 'alt text') for my Instagram posts to avoid inadvertently excluding anyone.

> There is a big trend on social media at the moment towards providing alternative text when you have text on pictures

10

Personal Websites and Personal Webpages

At the start of this chapter, I promised not to speak about web design. But I am going to finish by making a few comments that do stray into design territory. The first is to do with copyright of images, which we touched on earlier.

Before you paste up an image on your own website or that of an employer (if you have sufficient access rights to do so)—stop! Just take a moment to check that you have got copyright in, or a license to use, the image. From a legal perspective, and indeed a moral one, it is not OK to simply find a suitable image on the Internet and use it. But it is staggering how many people do exactly that. Unless explicitly specified otherwise, that image will belong to the person who created it. They may be a professional photographer, or a graphic artist, or simply someone interested in that particular subject. The point is that this isn't *your* image. All you have done is found it. If you did not create it, then you need to set about taking or making your own image, pay for the rights to use a desired photo or illustration, or search sites containing copyright free image content and choose one of those options instead.

When you are using other people's images with permission, make sure you have carefully read the usage rights of your chosen image and fulfilled any payments, crediting or other requirements for use. Not all licences allow for commercial use, for example, so you need to check diligently. Of course, it is better to get creative yourself if you can, so long as you don't infringe the copyright of anyone else's work by making a direct copy.

This may all sound like overkill, but I cannot overstress how important this is. When you paste content online you are a publisher. Yes, you! You are the person who has pasted stuff up for an online audience to see. So like any other publisher, you are responsible for ensuring that the content you publish does not infringe anyone's legal rights.

It may help to approach this in the same way that I do. Whenever I am due to paste or post anything online I have a mantra going round in my head that says: If in doubt, leave it out!

> When you paste content online you are a publisher

10

Communicating with Individuals and Small Groups

Don't Shoot the Messenger!

"Ideas like yours represent everything that's wrong with science communication in the UK!" fired the senior academic at me. This was after I suggested some potential projects during a meeting aiming to brainstorm new ways to encourage public interest in physics.

Even though I was shocked and quietly fuming inside, I tried not to take it personally, and explained the reasoning behind my suggestions in a calm but firm manner. I learned later that their outburst had put other people in the meeting off from airing their suggestions.

DOI: 10.1201/9781003206828-11

Clearly this negated the whole purpose of us gathering together in the first place! It really was a perfect example of how NOT to communicate with your peers.

In this chapter, we will look at some much more productive ways to communicate with fellow scientists, including tips on things to avoid doing.

Assuming Knowledge

One potential pitfall when addressing other scientists is sharing too much information. This is something Mario Merola, Head of the Engineering Design Department at the ITER Organization in France, finds he comes across quite often.

"One of the basic mistakes that I see when attending meetings with colleagues in the same area is a tendency to go into too many details. The psychological attitude is: what interests me for sure interests the others. So I need to explain why the size of bolt number 1000 must be 3 mm and not 4 mm, because this decision took me one week of work so they all should know why I've done that! This is one of the major mistakes because you can easily bore the audience, even if it is an audience in your same specialisation," he says.

Mario also advises against assuming that everybody is aware of what you are doing "which is usually never the case", and to avoid using acronyms. "I am really allergic to acronyms! Sometimes in a sentence there are more acronyms than real words. So I always say whenever you use an acronym please spell it out the first time you use it," he says, pointing out that it is usually only your immediate colleagues who will be as familiar as you are with these acronyms. "If you go just outside your close circle of daily colleagues your acronyms have no meaning and are not properly understood," he warns.

A good way to avoid such problems is to really think through what you are trying to achieve when you are communicating your work to others.

> Just outside your close circle of daily colleagues your acronyms have no meaning

"The first thing that you have to do is identify what your target is," advises Mario. "What do you want to achieve in your communication? Is it to improve awareness of what you are doing? Is it proposing a new research programme? Or is it reporting to your manager? And so on." Depending on what the objectives are and who your audience is, the communication should be tailored accordingly, he says.

"So for example if it is communication with your team to share awareness, then the focus would be on general aspects and trying to communicate in an attractive way. The slides should not to be too wordy—have less words and

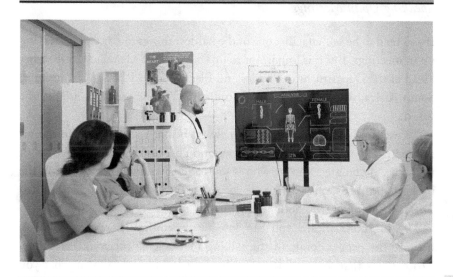

FIGURE 11.1
Tailor your communication to your audience and be clear on what points you are trying to get across. (Shutterstock ID:1918606010.)

11

more pictures to focus the attention where you want to. If you need some text, it's always good practice to highlight some keywords in bold or in different colours so they attract attention. Then you can develop your message according to these keywords." (See Chapter 6 for more advice on preparing clear-to-understand slides.) (Figure 11.1).

As well as aiming for targeted, attractive communication, it is vital to keep to the time available and not attempt to cram too much content into your time slot. "Use the golden rule of one slide per minute. Sometimes I see meetings in which a speech is supposed to last 15 minutes and the presentation is 50 slides. So of course either the speaker can never make it, or they only show a slide for 20 or 30 seconds so the audience cannot have the time to look at it and understand what it is showing," warns Mario.

It is also important to realise that different scientific disciplines can approach the same subject in a variety of ways. Dr Roger

> Use the golden rule of one slide per minute.

Fenn, an Emeritus Reader in Mathematics from the University of Sussex in the UK, has noticed this throughout his career when speaking with physicist colleagues. "A mathematician's attitude to mathematics is quite different to a physicist's attitude to mathematics. In general, physicists approach maths as a tool. They're not too bothered if what's under the bonnet is spluttering and backfiring as long as it gets you there [to the required answer]. Whereas a mathematician will want to cross all the t's and dot the i's," he says.

Speaking Up in Meetings

If you tend to be self-effacing, one of the hardest things to do can be speaking up in meetings. But it is important to do so, says Prof Makaiko Chithambo, Head of the Department of Physics and Electronics at Rhodes University in South Africa, who has worked and studied in various countries around the world.

"If we think of seminars, there will always be one or two people in a seminar room who will dominate the conversation because they are excited about the topic. If you are someone who is self-effacing you will tend to be in the background. You won't participate because you think you shouldn't do that. This is the wrong way of looking at things, I think."

Keeping quiet means you can lose out scientifically, warns Makaiko, because the people doing all the talking may not be correct all the time. "To counter that, be confident in what you know, and even in what you don't know so you can ask questions and find out. You have to admit what you don't know so you can ask: How do we solve this? How do we explain this?"

If you can't follow a train of thought, or a graph is not making sense to you, admit this, he advises. "By doing that you are expressing a lack of confidence

> Be confident in what you know, and even in what you don't know

qualitatively. But you are expressing it with an intention to learn rather than to display your ignorance, which are two completely different ideas" (Figure 11.2).

Makaiko feels it is also useful to bring this way of thinking to your own presentations, and not to worry about sharing speculative results at scientific conferences. "Your most important results are often the ones you can't immediately explain, and that you need to speculate about. Sharing your speculative explanations tends to invoke other explanations from the audience, so you might just get an insight from elsewhere. It comes back to being confident in what you don't know. You are losing nothing, because if a critique of your work comes with sensible reasoning it works in your favour. If you just want to have everything ticked off, sometimes you don't learn anything."

FIGURE 11.2
Take care not to lose out scientifically by keeping quiet in discussions. Ask questions to deepen your understanding. (Shutterstock ID:1936238746.)

11

Creating Clear Technical and Reference Documentation

Communication with your fellow workers is, of course, not restricted solely to talking. There are many scenarios in which you must document your work for reference by current or future colleagues.

For example, hardware engineer David Culpeck, explains that he has to accurately document several aspects of his work—which has included designing hardware for computer systems inside vehicles and on-board airliners—as an integral part of his day-to-day activities.

One of the required documents, all of which must be suitable for future use by colleagues who may need to maintain or develop the design, is a Hardware Design Description. "This describes the circuit function for a piece of hardware (electronics) and details any design limitations and assumptions, calculations and simulations used to produce the design," says David, who has over 30 years' experience of working on safety-critical projects. This description is kept with a schematic diagram of the design, and a list of the materials and components used. It is not uncommon, David explains, for problems to surface, or modifications to be required many years after the original hardware was produced and when the hardware designer involved is no longer at the company.

But while documents detailing the safety tests the hardware has passed are mandatory under regulatory requirements, David says that sometimes engineers can find the time required to create documents detailing the design gets squeezed. This, he says, is because of a constant push to have a fast time to market for new products.

"As mitigation for being asked to skip the design documentation stage, I tend to put as much information about design intent into the schematic [diagram] as possible to help those who come after [me]," says David.

Another pitfall for engineers when pushed for time is to go into "too much detail for the trivial aspects and not enough detail for elements that are highly non-standard," continues David, warning that whenever things are not standard, future colleagues "will definitely need guidance".

"With limited time, simply documenting the non-standard feature first is probably the best approach," says David, warning that otherwise hardware can become unmaintainable in the future due to the lack of complete documentation.

11

He also advises thinking carefully about any filing system you create on a computer, so that future colleagues can readily follow your thinking and get to the required

> With limited time, simply documenting the non-standard feature first is probably the best approach

documents. For example if there are revised versions of a design, David designates the first sub-folder layer as "Issue_1", "Issue_2", and so on, with each of these sub-folders then containing further sub-folders to house all the required documentation for each design.

By contrast with some of the filing systems he has inherited and struggled to navigate, which used number codes that he wasn't familiar with for projects or electronic components, David makes sure to label his folders using text. When possible this is a short description of what the folder contains, as whilst computers file quite happily under numerical codes "numbered information isn't that useful for humans!" chuckles David.

Dealing with Your Boss

Using code numbers can equally be a problem when relating information firsthand. A friend who manages several hundred staff members once moaned to me that people kept emailing to ask advice about specific components that they only described via their part numbers. My friend obviously could not retain thousands of code numbers in their head, so had no clue what the messages were about!

Although I've never had this type of role myself, it was easy to empathise. Neither my friend, nor anyone else, can be all over every tiny detail of a large project. But there are plenty of times when your manager or supervisor will need to make a decision about a certain aspect of your work. So in order to help them make the best decision possible, they will need clear information from you.

However, imparting this information can be easier said than done. Especially if you get nervous talking to more senior colleagues. To help combat any jitters, actor, facilitator, and communications trainer Gary Bates, who has worked for multi-national corporations including companies in the science and engineering sectors, stresses that it is important "to see everybody as a human being. Try to remember that everyone has insecurities, and doubts and worries and that generally 99.9% of the time people want you to do well. They want you to succeed."

"As a freelance actor you're in a position of no or very little power at all, and this is one of the things you have to keep telling yourself when you go into an audition room with a casting director and a director. It's very easy to see a director as the enemy—as somebody who wants you to fail. But actually these people want you to do well. A casting director has brought you into that room because they are recommending you for the role. So you have to think to yourself: 'these people want me to succeed'. That helps you to have inner confidence," says Gary.

11

"In business you've got your strategy and your business objectives and in the science world it's the same. You have an objective that you're working towards whether it's a piece of research, or you're trying to discover something, or whatever it might be. You're all working towards the same goal," he continues, adding that it is also important to try to relate to your boss on a more personal level.

"For me it's about human connection. The more you can know somebody, the better communication you can have with

> You're all working towards the same goal

them. When we work with managers and leaders one of the key things is about building trust, and building a relationship where you can have difficult conversations. Because things will go wrong in organisations and will go wrong in science I'm sure. It's not about getting to a state where nothing will go wrong. It's about getting to a state where when things do go wrong you are able to have an adult to adult conversation about it. You're much more able to do that if you have some relationship, some rapport and some mutual trust," explains Gary.

"You get that [trust] by doing quality work, being reliable, and delivering on what you're supposed to deliver on. But you also get that by having a human connection, by knowing the person. So don't be afraid to spend some time at the beginning and end of conversations talking to each other as people," he says.

"Try and find out something about the other person. Who are they? What do they like? What's their life situation? What makes them tick?," Gary advises. "I think that is a really important thing, because until we're all replaced by robots we are all people! Even the most senior, well-paid expert is still just a person."

> Don't be afraid to spend some time... talking to each other as people

Gary also advises altering your mind-set if you tend to get stressed or thrown by questions. "Don't be afraid of questions or a challenge. See them as a positive, rather than a threat or a negative. If someone is coming back with a question that's an opportunity for you to engage in a conversation with them," he states.

Subject Specialists

11

You may have noticed that as each rung of the career ladder is climbed, going higher up the scale of management, that those job positions will involve managing people with an increasingly wide range of specialties and skillsets. Clearly, no single manager is likely to have multiple PhDs in different science areas. Nor can they have experience of working in every role that they are now overseeing. This means that for many scientists the need to be able to explain their own project in an accessible and succinct way to their workplace superiors is crucial.

Hardware engineer David Culpeck recommends that when communicating verbally with managers who may not have the same specialist expertise as you, "do not go into too much detail. Try and pitch the technical content to their actual level of understanding of your field. Also, keep trying to gauge if what you are saying is actually going in." If at any time you don't feel you are being understood, David advises to slightly reduce the technical level.

By contrast, his advice for training more junior colleagues is to give a practical demonstration as well as providing a good written definition and some examples. This,

> Keep trying to gauge if what you are saying is actually going in

explains David, allows you to "talk through the 'what' and the 'why' as you go. As with managers, keep trying to gauge if what you are communicating is actually hitting the mark and adjust accordingly."

"[When training others] I also find asking them to explain back to me what we have just gone through is a good feedback loop to see if we are both on the same page at the end," adds David, who stresses the need for clarity no matter who you are speaking to in the workplace.

"Always use precise unambiguous language. For example, bullet point lists instead of long sentences with commas or semicolon punctuation will always work better to convey your meaning," he advises.

Scientific Rivalries

There is nothing like a bit of healthy competition between research teams to help move scientific understanding on at pace. But taking things too far, perhaps in the form of public slanging matches if you disagree on ideas, are best avoided, as is falling into the trap of having long, drawn-out arguments in print or online. (See Chapter 9 for advice on reviewing colleagues' books for newspapers and magazines.) One question you should definitely be asking yourself before getting into such a situation is whether having public spats is what you want your legacy to be?

Famously, the 17th Century English physicist Robert Hooke (1635–1703) was renowned for rubbing people up the wrong way. One of his many disputes occurred when Hooke attacked the first major scientific publication by the English physicist and mathematician Sir Issac Newton (1642–1727) in which Newton described his experiments on light and colour. This, coupled with several other spats between the two men, wound Newton up so much that he delayed the publication of his famous book *Opticks* until after Hooke had passed away.

11

Public slanging matches if you disagree on ideas are best avoided

In this day and age, it is unlikely that waiting until your rival has died before you publish is viable as an option. But there are plenty of ways to have rigorous debates in a productive mind-set that help science progress, rather than descending into an undignified cat fight.

It is also important not to engage in 'points scoring' over your fellow scientists or students in less public scenarios. People's memories can be long, and you never know when you might be required to work with that person again. Showing everyone the respect that you would like to receive from them— even if it isn't reciprocated—is not a bad motto to live by.

Classmates

We generally get plenty of practice communicating science while we are students. If nothing else, during our degrees we often have to collaborate with our fellow students on projects or assignments. It goes without saying that this is best accomplished by going about the task in an informal and friendly way. So if you get paired or grouped with someone who you don't normally get along with, treat this as valuable practice for needing to be able to work with anyone in your future workplace (Figure 11.3).

If you are lucky enough to be able to work with existing friends, that also makes good practice for future scenarios. For instance, a friend might end up promoted into a position of authority over you at work while at the same time being your weekly squash partner or picking your kids up from school. Practising wearing different hats in different scenarios, and agreeing to set any required boundaries between the two of you within the work situation, will stand you in good stead for the future.

11

FIGURE 11.3
Collaborating with fellow students during our college or university studies gives us useful practice for dealing with future workplace communication scenarios. (Shutterstock ID: 644329732.)

As with all communication scenarios, you cannot assume that every one of your classmates understands a topic in the same way that you do. So if you sense that you are not being understood, (politely) ask them if they would like you to clarify anything. By doing so you can hopefully help ensure that everyone in your project group is both metaphorically and literally on the same page.

Healthcare Settings

In some scientific roles, much of the one-to-one communication required will be with patients. The most important thing in this scenario is to "talk to patients like people and not objects," says Kathryn Adamson, a Principal Physicist in Nuclear Medicine and medical physics expert working for Guy's and St Thomas' NHS Foundation Trust in the UK. Her job includes having to explain the radiation safety precautions that patients must take when undergoing treatments that use radioactivity, some of whom will understandably be anxious about the procedures. Not least because these procedures use radioactive pharmaceuticals (given by injections or capsules) that remain in the body for several days, emitting radiation in the form of gamma rays. This necessitates Kathryn and her colleagues explaining to the patients the importance of minimising prolonged close contact with family and friends because of the risk this radiation poses. In the case of patients being treated for certain thyroid conditions, the radioactivity lasts longer in the body compared to other treatments, so the restriction of needing to keep away from other people has more impact, she explains.

"The types of questions these patients ask aren't necessarily technical ones, but they ask 'why?' when we tell them they

> Talk to patients like people and not objects

can't sleep with their partner for a fortnight," says Kathryn. While some people are fine with taking the spare bed, others question it. "They don't necessarily want a lot of scientific information, but they want to understand the reasons behind what you're saying," she continues (Figure 11.4).

There are instances of patients unwilling to go ahead with the treatment that day because of the restrictions they will have to follow immediately after. In those cases, Kathryn says, they can be referred back to their doctor. "In our role as physicists we're not there to persuade them to go ahead with their treatment. We are there to tell them about the precautions to take, as well as the benefits and risks, to enable them to make informed decisions."

Kathryn, who teaches about radiation safety and nuclear medicine to colleagues in related disciplines and to MSc and BSc students as part of her role, honed her communication skills via on-the-job training. She feels some of the

11

FIGURE 11.4
Some patients may not want to hear a lot of scientific information, but do want to understand the reasoning behind the things you are saying. (Shutterstock ID:1550930117.)

11

ability to put yourself into the shoes of a patient only comes with life experience, so is particularly challenging in the early stages of your career. "You can't teach a young person what it's like to be an older person undergoing cancer treatment. But if you fully understand your subject area, you can do your best to empathise and then engage in a conversation without worrying that you can't answer that patient's questions."

Although she says you need to be confident and knowledgeable in order to keep the confidence of the patient, if you are asked something beyond your expertise "the right thing to do is to say: I don't know but I'll find out and let you know." You should also be prepared for a wide variety in the amount of questioning. "Some patients may not want to know the details, but you still have a duty to tell them. Whereas other people may want to know a whole lot more factual information," explains Kathryn.

To ensure each patient can understand the science you are discussing, it is useful to ascertain as quickly as possible what level of scientific knowledge they have, she feels. "The key is to pitch what you're saying at the right level so that you're not patronizing anybody but you're not making it so complicated that they don't understand. You need to be friendly to put patients at their ease so they don't mind asking even what they might think you perceive as a stupid question. Those are the questions we want."

For many treatments, patients are sent away with information leaflets about the procedure, giving them a chance to digest more details later on. This, feels Kathryn, is important "because no matter who you are, you may only hear what you want to hear when you are in a stressful situation. You can either hone in on just the bad bits, or on just the good bits, and you can often forget the whole of what is said."

But no matter who the patient, or how they react, always talk directly to them, don't speculate on what questions they might ask, and be confident in your own subject area, says Kathryn. "If you can get your message across so the patients feel at ease and have not got any concerns, and maybe even leave with a smile on their face, then that's a good job done," she concludes.

The Public Face of Science

Of course, there are many different occasions when you may need to speak with members of the public. If communicating with the public is not a regular part of your job, you could still be required to do so on certain occasions. Many academics will want to, or be expected to, participate in face-to-face outreach activities. These could be in a setting such as a science festival, or during an open day at your university, institution, or company (Figure 11.5).

One of the most important things to remember is that meeting you might be the first encounter a person has ever had with a scientist. You will surely want to leave a positive impression, so it is crucial to tailor your content to the audience you will be addressing (advice on how to do this can be found in Chapter 3). It is also important to remember to be a human being! On undergraduate admissions days at the University of Sussex when I used to show groups of applicants' parents around, I was initially surprised by the types of questions they asked.

Although I had been tasked with demonstrating some of the physics equipment that their sons and daughters would potentially be using, the parents were often much more keen to discover how enjoyable the degree was than to find out more about its scientific content. They would ask how difficult the courses were, for example, and whether the lecturers were approachable.

To begin with, I had only prepared for answering questions about the science I was showcasing, and depending on the background of the parents on

11

FIGURE 11.5
You may get your first taste of communicating with the public at an open day for your university or institution. Be prepared to answer all manner of questions about science and its applications and implications. (Shutterstock ID:189094115.)

11

some occasions I would indeed be asked detailed technical questions. But equally there were days when I did not use any of my pre-prepared answers because all anyone wanted to know about was student life at the university.

This demonstrated two things to me. First, that you never have any clue as to what the knowledge level of your audience is when you are addressing the general public. Second, some of your audience members could be much more interested in the implications or applications of the science that you are discussing rather than its technical details. So be prepared to be asked all types of questions.

Be prepared to be asked all types of questions

12

Job and University Interviews

The Perfect Fit

12

"Take a seat," said my interviewer. I stared at the only option available and felt my anxiety levels rising. This was part way through one of my interview days for a place on an undergraduate physics course, and was an unwelcome development for my then 19-year-old self.

I made my way towards the chair that was positioned at right angles to the desk that my interviewer was by then sat behind. The room was so full of books and paperwork that it was impossible to move the chair. Worse still, it did not seem to have been designed with an office environment in mind as it was much lower than my interviewer's chair.

There I sat, trying to answer physics questions feeling like a small child as I peered over the piles of manuscripts on my physics interrogator's desk. It was also extremely uncomfortable trying to twist around in my seat enough to have any chance of maintaining eye contact with this academic, who at that juncture held my future career in their hands.

Somehow I managed to keep going with my answers, and I did get offered a place at that institution (which I declined for various reasons). But I can clearly recall the completely inappropriate furniture putting me off what I was supposed to be saying. It turned the interview into a bit of a battle for me. Aside from the psychological impact of feeling like a toddler, due to my disability the twisting around thanks to the positioning of the chair started to become painful after a while.

Although this was many years ago I still cringe at the recollection. But I also realise how much it taught me. Since then, if faced with a situation where furniture could put me in pain, or if the room layout is such that it is impossible to sensibly accomplish what is needed, I speak up. On several occasions in interviews for writing or broadcasting contracts I have politely asked for an alternative, explaining briefly why the provided furniture is hampering me from answering the questions to the best of my ability. (In all cases, people have willingly obliged.)

Having since heard many tales from both interviewers and interviewees, I realise that my teenage interview incident was minor compared with what can happen. Now I also acknowledge that it is awkward episodes like this which give us all the experience to deal with other tricky situations in job or placement interviews.

12 *Butter Fingers*

Software developer Rob Scovell, who has worked internationally for over 25 years including for multinationals, certainly changed his approach after attending a job interview that seemed like something out of a movie, but definitely did not follow a script that he would have chosen.

> Awkward episodes ... give us all the experience to deal with other tricky situations

"It was a very high powered interview in 1998 with the manager of a group doing mathematical modelling in a multinational company. My training was in mathematical and statistical modelling, and I'd learnt to use a software package called Genstat®, which not many people had skills in," recounts Rob, who has a bachelor's degree in astrophysics, and Open University Level 3 modules in computational methods for mathematical modelling.

At that time, he says, his approach to his career was to become highly skilled in something quite obscure. This had resulted in a call from an IT recruitment agency who had found Rob's CV on the internet, and asked to put him forward for a role at the multinational. "I'd done the training with the Open University, and they had a space [online] where alumni could put up their CV," continues Rob, adding that to be spotted online today he would recommend posting on GitHub if you're wanting to get into software development, or Kaggle if you are looking for data scientist roles.

"The IT recruitment agency flew me from Edinburgh in Scotland where I was living down to Gatwick in England. The head of the agency then picked me up from Gatwick airport and drove me to the research campus for the company," recalls Rob. He says that he had not been at all prepared for the five-star treatment he was being given. "I was incredibly nervous and stuttering all the time. I felt so out of my depth. I'd never been in anything like this situation before."

"I felt like a total impostor as well," continues Rob. "Because all I'd done was an Open University Level 3 course in the kind of stuff the company wanted me to do, using this particular statistical [software] package. They knew I had no commercial experience in this, but that I'd done some programming in my first degree and in my Open University courses. So I was thinking: 'This is crazy!'"

They duly arrived at a huge research campus, and the recruitment consultant dropped Rob at the gate. Once inside the campus, "I was treated like absolute royalty," he recalls. Just before his interview, a receptionist asked a still nervous Rob if he would like a cup of coffee. He accepted the offer, but almost immediately regretted that decision.

"I was called in to meet the interviewer and I fumbled the handshake. I was so incredibly nervous and shaky that as I sat down the cup of coffee went all over me. I was wearing my best suit and my best shirt," recalls Rob, revealing that as well as ruining his attire the coffee also spilt onto the interviewer's desk.

Rob was understandably mortified, and to this day worries that some of the coffee went into paperwork on the desk. But fortunately, this incident did not faze his interviewer.

"He just looked at me and said: 'Oh, that's unfortunate.' Then he carried on as if nothing had happened! He asked me a few questions about statistics and if I used Genstat, and if I knew [computer programming language] Fortran and how to use Excel and Word. He then chatted and showed me around the campus."

12

"I'm from a fairly humble background and this [interview opportunity] was via a very high class recruitment agency, which had added to my nervousness. I felt completely like a fish out of water," continues Rob. His feelings of not belonging were compounded during the journey to the research campus, when the head of the recruitment agency had offered Rob to share travel in their chartered private jet if he decided to commute weekly from Edinburgh to Gatwick. "I thought: 'Oh my goodness what is this!'," he recalls.

As it turned out, Rob needn't have worried about his nerves and consequent clumsiness because he was offered the job. Even if he had not been, he feels the experience was a valuable one as it taught him that once he got talking about his area of expertise, his interview nerves would disappear.

"If you're the kind of person who is nervous, but loses that anxiety when talking about things you know there's no shame in going straight into the geeky stuff that you are being interviewed for. Get into your 'nerdzone' as quickly as possible because that's what is expected of you. Don't feel pressure to be a social person if that's not your strength," advises Rob.

"If you've got to the interview stage because you are highly skilled in something highly specialised, it is not going to be unexpected if you go into nerd mode

There's no shame in going straight into the geeky stuff

about the thing you're being recruited for," he continues. In the case of his 'coffee incident' interview, Rob later discovered that not only had he been the only candidate with his specific background of expertise, but also that he had something in common with the interviewer.

"When I got to know that interviewer afterwards, I discovered that we are both geeks and that he was also very clumsy! I think the coffee spill was completely normal for him. It was the most bizarre scenario I'd been in up to that point in my life. But it had calmed me down that he wasn't fazed," recalls Rob, who in addition to rapidly drawing on his inner geek has made one other permanent change to his interview technique as a result of this experience.

"I don't ever risk having a cup of coffee, or anything else. Even a glass of water is a potential disaster area!" he chuckles (Figure 12.1).

What to Expect from a PhD Interview

For many people working in science, their first experience of an interview will be for a PhD position rather than for a job. The sorts of questions you get asked will differ depending on the type of PhD on offer, says Professor of

FIGURE 12.1
Don't be afraid to move quickly into talking about the technical stuff that the job involves if that helps reduce your interview nerves. If you have a tendency to be clumsy, consider turning down any offers of hot drinks like coffee. (Shutterstock ID: 1996276739.)

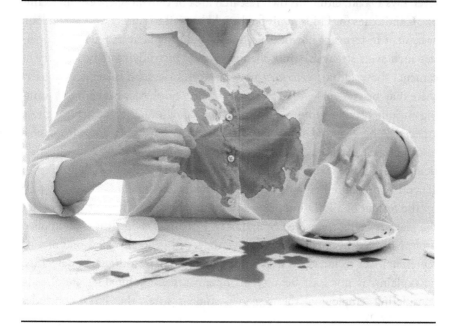

Physics David Faux from the University of Surrey in the UK, who has conducted many PhD interviews.

"If it's in a core area of physics for example, such as nuclear physics, as the interviewer you would be looking at what the candidates had done at undergraduate level. You would expect to see that they had taken optional nuclear physics modules or done a project within that field. Then their project or modules would be your focus for questioning."

"As the interviewer you're trying to gauge the genuine interest of the applicant in the area of the PhD under offer," continues David. So if they've selected optional modules in that area then that demonstrates their interest, he explains. "You would want to quiz them about the content: what they had learned from it, and what they found most interesting. You would also get the student talking about their project. But it would be quite general questions. It's not a test. It's about getting the candidate talking about the subject, and

12

gauging their enthusiasm and understanding, and if they are truly motivated to do the PhD."

By contrast, there are some multidisciplinary PhD subject areas where you wouldn't expect students to have encountered the research topic at undergraduate level, says David, citing his own research area of cement science as an example. Candidates for PhDs in his field could be civil engineers who would look at cement in terms of how it is used in buildings and at its physical properties such as its strength, while chemists would be studying the chemistry of cement—how adding water to a powder creates the hard substance. Equally there may be students with a materials science or physics background wanting to join David's group who would not know much about any of those aspects. Where candidates can't be expected to have prior knowledge of a research area, David says he will ask them why they've applied for this particular PhD to see if they have some understanding of the discipline. "But in general I am looking for a desire to do research," he states.

> It's about getting the candidate talking about the subject, and gauging their enthusiasm and understanding

The format of a PhD interview will mainly be dictated by whether or not the place is being sponsored by industry, explains David. If it is an applied project funded by a company he says they will generally send one or two representatives to interview a selection of candidates on the same day. This, he explains, is not least because in some cases the companies are already considering employing the research student after the PhD concludes (Figure 12.2).

For PhDs being offered by university departments, it is more likely that you'll be invited to attend the interview and be shown around laboratories and other facilities on campus separately. This latter type of interview will probably include more one-to-one interaction with your prospective supervisor, and also tends to be more informal, says David.

How stiff the competition for a place is will depend on the research area. PhDs with more money available for the candidates tend to be popular, as are certain research areas such as quantum technology and astrophysics. But even when there are fewer applicants, landing a PhD place is not easy. So you always need to show yourself in the best light possible.

How to Avoid Putting Off Your Potential Supervisor

If you've got a really good mark for your project work, there's no harm in bringing the report along, but leave it at home if your grade wasn't high, advises David, who admits he was once put off a candidate who showed him what looked more like a draft than a completed project report.

FIGURE 12.2
When PhD studentships are sponsored by industry, the company will likely send one or two representatives to interview a selection of candidates on the same day. So expect to find yourself meeting other candidates, and possibly touring facilities with them. (Shutterstock ID: 447401257.)

It would also put David off a potential PhD student if they were unclear on what they were doing in their project. "You need to be on top of what you did and need to be able to explain it to people who don't know some of the science. You also need to explain your motivation for that particular project, and to describe clearly and succinctly what you did and how you did it, and what the key outcomes and overall conclusions were," he says.

In advance of your interview, David advises to "prepare a two-minute summary of what your project was about, and practise this by finding a friend or relative and asking them to listen". This, he explains, helps you to hone your explanation into something readily understood.

> You need to be on top of what you did

After you have delivered this summary during your interview, you are then likely to be asked follow-up questions about your project, so be prepared to explain the science in more depth then, says David. "[As an interviewer] you want to find out how much they really understand about what they did.

The candidate is the expert in their project and they need to show it. Nobody knows their project better than they do."

However if you really can't answer a question, "don't try to bluff and bluster," advises David, stressing that it regularly happens that academics are asked questions about their research areas that they cannot answer. Instead he recommends saying something like: I really don't know—that's an interesting question! Just talking around the work topic will still show your interest and enthusiasm, he feels. "Interviewers will appreciate that a final year project is time limited and that you can't explore every avenue. So you can engage in some sort of conversation even if you haven't got a clue what the answer to a specific question is."

Other things to avoid in a PhD interview are coming over as bored or being passive, says David. "You want someone who is passionate, enthusiastic and highly motivated for the PhD project. Three years is a long time to be working on something. You want someone who really wants to do this work, and someone who you can talk to and engage with on any topics—even those outside of research. As a supervisor you want to feel as though you can get on with this person for three years."

It's All about You

Equally, as the applicant you need to be sure that it's somewhere you would be happy working. So David recommends you take the opportunity to ask the interviewer questions, particularly while they're showing you around the campus or research facilities. "There are usually lots of opportunities for informal discussions," he says, suggesting that you ask questions such as when they started working in this area, and what they find most motivating about their work.

"Be interested in them because you are trying to find out what it is like to do a PhD with this person in this subject area. Talk to some of their current students too. Ask them what it is like to do a PhD there, and what part they find the most difficult." Those sorts of conversations are important, feels David, not least because you must also be able to successfully engage as part of that team.

It is worth taking a similar approach in job interviews, which are not just a chance for your potential employer to hear what you have to say. They are also an opportunity for you to assess whether or not you want to work for that employer. So make sure to ask anything that you need to in the interview.

Find out what it is like to do a PhD with this person in this subject area

FIGURE 12.3
Before you accept any job offer or PhD placement, make sure that you are happy to work in that environment by asking the questions that are important to you during the interview. (Shutterstock ID: 1707285751.)

Also do not hesitate to request a tour around lab facilities or to meet the team members who you will be expected to work with. Trying to gauge whether you feel a good fit with the institution or company is just as important as other aspects of the job role such as the salary and benefits (Figure 12.3).

12

Troubleshooting

I'm Stuck!

No matter how much any of us prepare, there can be last minute hitches that cause issues. So in this chapter, I have collated together some advice on how to handle the sorts of emergencies that can throw us off course completely when it comes to science communication.

Grounded in the Airport or at the Station

This is where all that practice with using remote technologies comes in! If you have transferred your talk slides to the host beforehand, and have a mobile

13

13

DOI: 10.1201/9781003206828-13

Some forward planning
can allow you to deliver
that talk even when stuck
at an airport or station

phone signal and enough charge and credit, you are good to go. None of us would choose to give a presentation from a departure lounge. But thanks to modern technologies, we could do this if absolutely pushed (Figure 13.1).

Another way around potential issues with travel, if you know there could be a problem with you getting to a venue on time, is to pre-record the presentation. That way you can send the video to the host, and they can play this out to the audience at the allotted time for your talk. You can then join the meeting for the Q&A session afterwards via video conferencing software.

I used this method once when I was having issues with my broadband cutting out. I was 'live' for the introduction, then my pre-

Pre-record the presentation

13

recorded talk was played out by the conference hosts, before I came back in front of my webcam again for the questions after. It worked fine, and got rid of a lot of technical worries for both myself and the host. Not every host may be willing to do this, but if you are travelling tight on a scheduled meeting session, or having any technical difficulties with remote connecting due to power cuts, broadband issues, or anything else, this may provide a solution.

My Broadband Has Gone Down in the Middle of Giving a Presentation

If it does not look like you will be able to recover the connection within a few minutes, there is only one thing for it. Get on the phone to the host! They can then hold the speaker of their phone to the microphone in the lecture hall (or other venue) and the audience will still be able to hear what you are saying. So make sure that you have a mobile phone number for the session host ahead of time.

With your slides supplied beforehand, these can be shown and moved through on your request by someone at the host's end.

> Make sure you have a mobile phone number for the session host

This Presentation Is Critical to My Future Career and I Am Panicking about It. How Can I Keep Calm Enough to Convey the Required Information?

The main thing you need is to minimise the impact of fear of failure, says sports psychologist Prof Richard Keegan, Professor of Sport, Exercise and Performance Psychology at the University of Canberra in Australia, who also works with international sportspeople and business clients advising on performance.

"Motivationally, if someone is focussed on avoiding something undesirable you can monitor their body and see it acting differently in terms of cortisol, heart rate and muscle tension, compared to when they are carrying out the same task but are motivated to work towards a desirable goal instead," he explains. So altering your mind-set and focussing more on what information you want to impart can literally alter your body's physical response to exactly the same presentation scenario, and thereby help to keep you calmer.

It is also important to rationalise your fears. "To reduce this fear of failure, think about what the consequences are if your talk goes abysmally. Say to yourself: 'How bad can this be? It's not life or death and no-one is going to be physically harmed.' It might be mildly embarrassing, but it's not as bad as your body is telling you," says Richard.

"Then contemplate what could go wrong but be realistic. For example, there's not going to be an alien invasion!", he continues, adding that then

13

putting plans in place to deal with any issues also increases your confidence levels. Knowing you're well prepared for tricky situations should clearly boost confidence. You can then either physically rehearse that plan, or rehearse it in your imagination." (See Chapter 6 for more advice on planning for mishaps during talks.)

That said, if you have a worry that just won't go away—especially if it nags at you during your presentation—don't try to block it or ignore it, advises Richard, make a mental note to jot it down in a 'worry pad' instead.

> Contemplate what could go wrong but be realistic

"If you can't deal with a worry immediately, write it down and—importantly—make a date to return to it and review it later. Supressing and blocking worries famously tends to backfire, but allocating them a specific time to be heard can calm the urgency and let you continue to perform. When people do return to review these repetitive worries, they have often resolved themselves by then, illustrating that we shouldn't have given them so much attention," he says. This also means that you can quickly sidestep the distracting thought, and press on with answering questions or giving your presentation because you know that you've already set aside time for dealing with it later on.

Deciding beforehand on the main message for your presentation should also help prevent things running away from you when you deliver it. "I always think the intent can be my compass," says Richard. "If I feel a bit lost, and the room is looking a little bit blank I've still got that compass going: What was I here to achieve? And if I go back to that I know I will be OK. The point of a presentation is not to improve your self-worth, so you shouldn't make it about that."

I'm in the Middle of My Talk and Feel I Just Can't Go On. What Should I Do?

> The point of a presentation is not to improve your self-worth

13

It's important to know when to quit, and simply to cut your losses. "I think having a really clear mental picture before you walk in the room under what circumstances you just can't carry on is actually helpful," says sports psychologist Prof Richard Keegan from the University of Canberra.

This is especially useful in situations when technology fails. "Say things get pushed so far behind schedule that you haven't even started speaking and you only have five minutes left before the tea break. If there's no chance of running into the tea break you could just say: 'I wish that hadn't happened, but I can record the information and I'll send it over to you as a video clip afterwards'." He feels it is important to remember that the purpose of the session is to get over a certain amount of information, or a key message, so if you have to do that in a different way after the allotted time slot it's not the end of the world.

Nevertheless, a lot of technical difficulties can be planned for, and Richard advises making sure you've got built-in redundancy for crucial meetings. This may take the form of having a printout of your talk or even taking a spare computer or a second notebook with the same information in.

If nerves start to get the better of you, you may also need to pause or quit your presentation and send information afterwards. "If you feel unable to formulate words, can feel your heart pounding out of your chest, or you're having a panic attack, have a rehearsed phrase that you can use to save face and then exit," advises Richard, stressing that this judgement can be tricky to make.

"In some situations, such as in the middle of giving a presentation, I don't think you have the insight to know that this might be a panic attack. Important parts of your brain are basically being starved of processing power and in some cases even getting reduced blood flow. So that's when you either want to have a phrase thoroughly rehearsed, or the ability to press pause and take a breath."

"Most of us will have experienced a panic attack at some level," continues Richard, admitting that when he started to go into one during an important meeting he excused himself for a few minutes by saying he felt unwell and wanted to take a short break. On his return, he successfully completed the difficult negotiation, ultimately salvaging a half-million dollar contract.

So if your physical symptoms are becoming such a problem that you fear you might just break down in front of the audience, Richard feels it is much better to just explain—quickly and simply—that you feel unwell and that you need to take a few minutes out. However, he stresses that it is highly unlikely you will need to do this if you have carefully thought through some of the things that could go wrong during your preparation and have strategies in place to mitigate.

More minor attacks of nerves, like feeling butterflies in the stomach can be a good thing according to Richard because these types of nerves show that you care about your performance. Many elite athletes interpret these symptoms as signs of being ready, not of impending doom. He also recommends lowering your expectations away from demanding perfection so that you are not devastated if your presentation is anything less than perfect.

"This change in perception can be extremely useful. One gymnast that I used to work with said that one of the things that improved their performance was to expect things to go wrong during major competitions, such as the organisers playing the wrong music for their floor exercise. This gymnast decided they would defer having a tantrum until at least five things had gone wrong! This change in their expectation, just accepting that things won't be perfect, really helped them. Oftentimes, making our bad performance better has more of an impact than extending our 'top end' capability: you want to have 'good bad days'."

Richard advises taking a similar approach with any public speaking that you need to do. "Change from thinking in black and white to shades of grey, and know when to quit if you're at risk of making it into a bad experience."

Going into a Total Panic the Night Before

Well, if it is any consolation, we've all been there! The most nervous I have ever been was giving the presentation for my third-year undergraduate project at university. I was beside myself with nerves.

Speaking to a friend in the entertainment industry the night before only increased my nerves as the more I chatted about it, the more there seemed to be at stake. If I messed this up, I would get a poor grade.

On the day, I confessed to a classmate how nervous I was feeling, and asked if they would be good enough to grin at me from the back of the audience. They duly gave me an almost cartoon-style over-the-top grin half way through my talk, which frankly did not help much as it almost made me burst out laughing. The latter would not have been at all appropriate! So I now had the added problem of needing to stifle the urge to giggle.

Somehow I got through the talk, and I can recall getting a good grade for it. But in all honesty this was so long ago now that I can't remember exactly what grade I did receive. And that is part of the point. What seems like a pivotal moment is in reality likely to be one of a whole series of moments you will have throughout your career, none of which individually are likely to have the devastating effect that you are imagining right now. No one will remember years later if you stumble over a couple of lines, or realise part way through that you've left something out and need to add it in a bit later.

13

What people recall is the overall impression and (hopefully) a few of the standout facts that you are sharing. We all go into a conference session, lecture, or public talk wanting it to be good. So that goodwill is there from your audience.

> No one will remember years later if you stumble over a couple of lines

Researchers will all want their research area in general to progress, and will want to hear about new results. They may have in mind a potential collaboration, or might want to build on their own work. When it comes to public science talks, we all want to be entertained as well as learn something new. So always bear in mind that it is not in the interests of your audience members for you to do badly, which means they are highly unlikely to give you a hostile reaction.

Feeling Like You Don't Belong

For some people, the thought of getting up in front of others to speak, or seeing something they have written in print, will always be daunting no matter how much they prepare their material and themselves. This might be due to feelings of inadequacy, or impostor syndrome, or believing that you just don't 'fit in'. These types of thinking can lead some people to fear they will fail at whatever task they are trying to accomplish because that task is 'not for them'. Please don't be one of those people! There is a place for everyone within science communication (Figure 13.2).

Often when I've felt daunted by a new situation I think back to a particularly memorable school sports day. As the only

There is a place for everyone within science communication

child in the class with a physical disability, I used to dread sports day. All that seemed to matter was how fast you could run or how high you could jump, and this appeared to be directly proportionate to how friendly many of my classmates would be over the next school year. There wasn't much I could physically manage to do, but I was entered for the egg and spoon race. (For anyone who hasn't had the pleasure of taking part in one of these races it involves running along with a hard-boiled egg balanced on a wooden spoon held out in front of you.)

We all lined up in the various running lanes. Then we were off! I had very low expectations of myself. I assumed that due to being so slow at running compared with my able-bodied classmates that I would be certain to come last, and no doubt be laughed at as a result. At first it went as I'd predicted, with fellow pupils shooting past me. But one by one their pace proved too fast for their eggs, delicately balanced as they were on their spoons. Off the eggs fell, and as my competitors were stopping to pick them up I began to gain ground. Of course I could never hope to directly out-run them. But unlike my classmates I was being really careful with my egg. It never once dropped off the spoon.

FIGURE 13.2
Never feel like you don't belong—you do belong.
(Graphic: Canva. Text © Sharon Ann Holgate.)

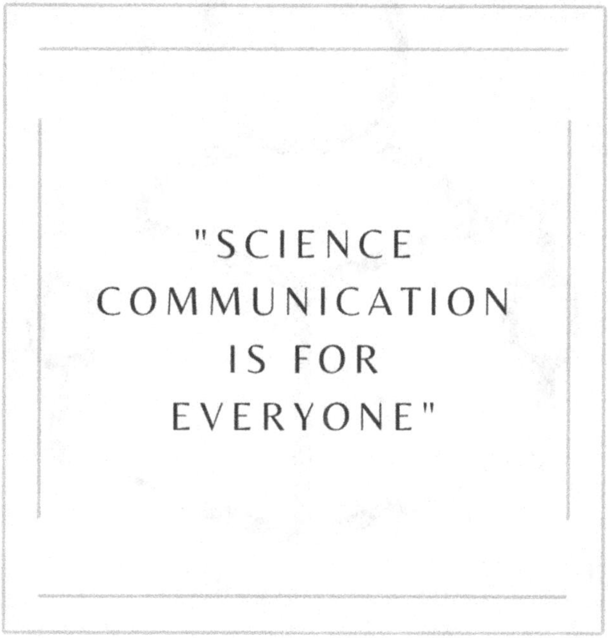

"SCIENCE
COMMUNICATION
IS FOR
EVERYONE"

Had I kept going at this stage, I would have won the race. But I could not believe I was out in the lead and I stopped running and turned around in astonishment at my progress before I had reached the finish line. My classmates were grabbing at the eggs strewn around the grass and with the precious cargo back on-board their spoons, quickly setting off again. Sure enough, because I had completely stopped moving two classmates were able to overtake me and I came home in third place.

As I got older, I took this incident to be a metaphor for life. I haven't stopped and looked round since. And neither should you.

THE END—but hopefully not the end of your science communication journey!

13

Further Reading

In order to keep the references as up-to-date as possible, instead of listing them here, my Further Reading suggestions can be found as a downloadable pdf from the CRC Press website page for *Communicating Science Clearly* at http://www.routledge.com/9781032069111

By doing it this way, I can update the pdf periodically, and so provide you with a more useful resource.

This pdf also contains links to the videos I have made to accompany this book, which give additional science communication tips and advice. As with the references to the online articles, there are smartphone-compatible QR™ codes alongside these links that will take you directly to the video content.

Index

Note: *Italic* page numbers refer to figures.

Printed in the USA
CPSIA information can be obtained
at www.ICGtesting.com
LVHW020608170924
791293LV00001B/212